BIOCHEMISTRY RESEARCH TRENDS

MICRONUCLEUS ASSAY

AN OVERVIEW

BIOCHEMISTRY RESEARCH TRENDS

Additional books and e-books in this series can be found on Nova's website under the Series tab.

BIOCHEMISTRY RESEARCH TRENDS

MICRONUCLEUS ASSAY

AN OVERVIEW

ROBERT C. COLE
EDITOR

Copyright © 2020 by Nova Science Publishers, Inc.

All rights reserved. No part of this book may be reproduced, stored in a retrieval system or transmitted in any form or by any means: electronic, electrostatic, magnetic, tape, mechanical photocopying, recording or otherwise without the written permission of the Publisher.

We have partnered with Copyright Clearance Center to make it easy for you to obtain permissions to reuse content from this publication. Simply navigate to this publication's page on Nova's website and locate the "Get Permission" button below the title description. This button is linked directly to the title's permission page on copyright.com. Alternatively, you can visit copyright.com and search by title, ISBN, or ISSN.

For further questions about using the service on copyright.com, please contact:
Copyright Clearance Center
Phone: +1-(978) 750-8400   Fax: +1-(978) 750-4470   E-mail: info@copyright.com

NOTICE TO THE READER

The Publisher has taken reasonable care in the preparation of this book, but makes no expressed or implied warranty of any kind and assumes no responsibility for any errors or omissions. No liability is assumed for incidental or consequential damages in connection with or arising out of information contained in this book. The Publisher shall not be liable for any special, consequential, or exemplary damages resulting, in whole or in part, from the readers' use of, or reliance upon, this material. Any parts of this book based on government reports are so indicated and copyright is claimed for those parts to the extent applicable to compilations of such works.

Independent verification should be sought for any data, advice or recommendations contained in this book. In addition, no responsibility is assumed by the Publisher for any injury and/or damage to persons or property arising from any methods, products, instructions, ideas or otherwise contained in this publication.

This publication is designed to provide accurate and authoritative information with regard to the subject matter covered herein. It is sold with the clear understanding that the Publisher is not engaged in rendering legal or any other professional services. If legal or any other expert assistance is required, the services of a competent person should be sought. FROM A DECLARATION OF PARTICIPANTS JOINTLY ADOPTED BY A COMMITTEE OF THE AMERICAN BAR ASSOCIATION AND A COMMITTEE OF PUBLISHERS.

Additional color graphics may be available in the e-book version of this book.

Library of Congress Cataloging-in-Publication Data

ISBN: 978-1-53616-678-1

Published by Nova Science Publishers, Inc. † New York

CONTENTS

Preface		vii
Chapter 1	Clinical Applications of Micronucleus Assay *Merve Bacanli*	1
Chapter 2	Micrunucleus Test Evolution *Alejandra Hernández-Ceruelos, Sergio Muñoz-Juárez, Fernando Vázquez-Rivera, Jesús Ruvalcaba-Ledezma and Luilli López Conteras*	15
Chapter 3	Pivotal Role of Micronucleus Test in Drug Discovery *Hasan Türkez, Mehmet Enes Arslan and Adil Mardinoğlu*	49
Chapter 4	Micronucleus Assay in Occupational Toxicology Studies *Hatice Gül ANLAR*	75
Chapter 5	An Overview on Micronucleus Assay Evaluation by Flow Cytometry *María Sánchez-Flores, Eduardo Pásaro, Blanca Laffon and Vanessa Valdiglesias*	97

Chapter 6	*In Vitro* Micronucleus Assay: Scope for Genotoxicity Assessment and Beyond *Abhipsa VF Debnath, Priti Mehta and Sonal Bakshi*	**123**
Chapter 7	Study of Micronucleus in Bone Marrow Cells of Mice: A Review *Mehnaz Mazumdar*	**145**

Index   **161**

Related Nova Publications   **167**

PREFACE

Micronucleus is defined as the small nucleus that forms whenever a chromosome or its fragment is not incorporated into one of the daughter nuclei during cell division. It is concluded that micronucleus assay can be used for risk prediction, screening, diagnosis and treatment of various chronic diseases. In this Micronucleus Assay: An Overview, the applications of micronucleus assay will be discussed.

Ionizating radiations, ultraviolet rays, geogenic or anthropogenic pollutants can induce mutagenic, teratogenic or carcinogenic effects due to the induction of micro or macrolession over DNA. Several models have been used to measure the mutagenic and clastogenic effect of such agents. As such, the authors focus on one of these models: the micronucleus test.

The micronucleus test can assess abnormalities earlier in the drug discovery pipeline, making structure/genotoxicity connection a possible system for drug characterization.

Additionally, the authors provide knowledge about micronucleus assay and its usage in occupational toxicology studies. It is now recognized as one of the most successful and reliable assays for genotoxic carcinogens.

The authors go on to present an overview of the evaluation of micronucleus assay by flow cytometry, reviewing the studies published in the international literature so far that employ different experimental designs for a variety of purposes.

Humans can become exposed to a variety of chemical substances that can have adverse biological effects, and the sub-lethal genotoxicity can have the most far reaching and severe consequences like cancer or abnormal progeny as per the cell type involved. Hence, the penultimate chapter focuses on the significance of identifying and predicting potential genotoxic agents by using laboratory markers, thus regulating and preventing exposure to cancer causing agents.

In the concluding review, generation of micronucleus assay in the bone marrow cells of mice induced by various clastogenic chemicals, drugs, and radiation are elaborately elucidated. Bone marrow cells are easily susceptible to oxidative damage and sensitive to various clastogenic as well as aneugenic agents.

Chapter 1 - The cells in the human body are exposed to various endogenic and exogenic toxic substances which have the potency to cause genotoxicity. Genotoxicity may result in chronic diseases such as diabetes, cardiovascular diseases, Alzheimer disease, cancer and aging. Because of this, the detection of genotoxicity is very important. Micronucleus (MN) is defined as the small nucleus that forms whenever a chromosome or its fragment is not incorporated into one of the daughter nuclei during cell division. It is concluded that MN assay can be used for risk prediction, screening, diagnosis and treatment of various chronic diseases.

In this chapter, the applications of micronucleus assay will be discussed.

Chapter 2 - Ionizating radiations, ultraviolet rays, geogenic or anthropogenic pollutants can induce mutagenic, teratogenic or carcinogenic effects due to the induction of micro or macrolession over DNA. Several models have been used to measure the mutagenic and clastogenic effect of such agents; one of them with many advantages is the Miicronucleus test (MN). This assay can be performed *in vitro* and *in vivo*, using vegetal or animal tissue. The test can be performed at a low cost and, in most cases, it does not require highly especialized equipment. The test can be performed under controlled laboratory conditions or as a biomarker of environmental exposure to genotoxic agents. Micronucleus (MN) are defined as intracitoplamatic chromtine fragments separated from the main celular nucleus. The fragments are the result of chromosomal break or chromosomes that had suffered delay during anaphase, then MN has been recognized as biomarker of chromosomic damage. Therfore, the aplication of the MN assay has demosntrated a huge capacity to determine clastogenic or aneugenic effects under different conditions. First studies were performed in mouse bone marrow using histological stains looking for abnormal nucleus in an atempt to replace metaphases analysis. Later the test was performed in mouse pheripherial blood, but MN can be observed also in nucleated cells. In lymphocites cultured with cytocalacine B, as a citokinesis blocker, to obtain binucleated cells can be use as an *in vitro* test to observe micronucleus and other nuclear abnormalities. Another *in vitro* model is the culture of *Tradescantia paludosa* influorescences. Nowadays, the study of exfoliaed ephiteial cell for oral cavity or urinary epithelium has been used for biomonitorization in human population using the cytome model where apotosis induction can also been measure, as well as, binucleated cells and nuclear buds as indicators of chromosomal inestability.

Chapter 3 - Early detection of adverse effects of novel compunds during drug discovery and development most probably reduce late stage failures, expenses and exertions for candidate drugs. Although the micronucleus (MN) test is one of the oldest techniques used in biochemical sciences for drug discovery. Flexibility of the technique for both *in vitro* and *in vivo* applications and practicability for large scale samples in short time make the MN test an inevitable tool for chemical trails. Drug studies require a formulation that provides the highest exposure to detect clastogenic and aneugenic activities and thus analysis makes it possible to get the necessary safety margin to support clinical trials. The MN test is one of the most important tools of the genotoxicity test battery in preclinical studies to identify negative effects of compounds that induce numerical and structural chromosome alterations in wide spectrum concentrations. The MN assay can be applied various cell types in different protocols. For instance; the most recommended protocols are bone the marrow micronucleus analysis and the *in vivo* mammalian erythrocyte precursor assay. Also, the rodent ovary cells validation test is a very powerful approach to analyse side effects of a compound. Beside cell types, detection systems can be constituted to obtain a high throughput screening such as integrating flow cytometry analysis into the MN inspections. Since a new compound is needed for such an assay, the MN test can assess abnormalities earlier in the drug discovery pipeline, making structure/genotoxicity connection a possible system for drug characterization.

Chapter 4 - The micronucleus (MN) was recognized at the end of the 19th century when Howell and Jolly found small inclusions in the blood taken from cats and rats. The small inclusions, called Howell-Jolly body, are also observed in the erythrocytes of peripheral blood from severe anemia patients. After that, the MN test is widely used in toxicological studies and now recognized as one of the most successful and reliable assays for genotoxic carcinogens, i.e., carcinogens that act by causing genetic damage. There are two major versions of this test i.e., *in vivo* and *in vitro*. This test also widely used in occupational toxicology studies. A person spends, on average, one/third of his/her life at his/her workplace and therefore the environment in which he/she works can be a major factor in determining health status. Many studies have confirmed that the number of MN have increased in workers exposed to inorganic lead, painters exposed to lead-containing pigments, ceramic dust, polycyclic aromatic hydrocarbons, coal dust, and welding fume. MN can be evaluated in different kinds of cells that do not necessarily have to divide *in vitro* such as epithelial cells, thus, the analysis of MN in exfoliated buccal cells has been demonstrated to be a sensitive method for monitoring genetic damage in human populations. Since epithelial cells can be obtained easily by relatively non-invasive methods and are capable to indicate toxicity in actual target tissue by the MN assay, their usage in case-control studies has been increasing. This chapter aims to provide knowledge about MN assay and its usage in occupational toxicology studies.

Chapter 5 - Micronucleus (MN) assay is a commonly used method to evaluate chromosome alterations. MN are expressed in dividing cells as the result of chromosome fragments or whole chromosomes that lag behind during anaphase. Therefore, MN assay provide a reliable measure of both chromosome breakage and chromosome loss, and thus MN frequency is a widely accepted biomarker of genotoxicity and genomic instability. Traditionally, the method most commonly employed for MN assessment is microscopy. However, this technique is time consuming, highly subjective, and the number of cells scored is relatively low. In this context, flow cytometry, as a high throughput alternative method, allows the automation of the MN scoring to overcome the mentioned limitations of the standard microscopy scoring. Besides, in the last years, several imaging flow cytometry platforms have been developed to evaluate the MN assay, which enable the capture of high resolution images in addition to the traditional flow cytometry features. This chapter presents an overview of the evaluation of MN by flow cytometry, reviewing the studies published in the international literature so far that employ different experimental designs (*in vitro*, *in vivo*, human biomonitoring) for a variety of purposes (radiation biodosimetry, risk assessment, aneugen and clastogen classification, etc.).

Chapter 6 - Humans can become exposed to a variety of chemical substances that can have adverse biological effects. Among various types of toxicities the sub-lethal genotoxicity can have the most far reaching and severe consequences like cancer or abnormal progeny as per the cell type involved. Hence it is of great significance to identify and predict potential genotoxic agents by using laboratory markers and thus regulate and prevent exposure to cancer causing agents. The genotoxicity is majorly exerted through clastogenicity where the broken chromosomes with or without a centromere segregate separately from the main nucleus following cell division, called micronucleus. Thus it is considered as a surrogate marker for the chromosomal breakage, spontaneous or induced due to the exposure to an agent. The search for an ideal biomarker for genotoxicity that is robust, sensitive, and objective has resulted in development and optimization of lab assays like chromosome aberration assay, micronucleus assay, comet assay, gamma H2AX assay etc. The micronucleus assay being one of the promising bioassays, the authors aim to discuss in detail the strengths and shortcomings along with the applications and scope. Following acute or chronic cellular exposure to various agents the sub lethal genetic damage manifest as various structural and numerical chromosomal aberrations which can be best detected and quantified at cytogenetic levels. The fate of certain structural aberrations observed at metaphase is such that these can be detected at interphase stage in the form of micronucleus. The scoring of frequency of such cells *in vitro* and in vivo allows for larger sample size and automation as compared to the conventional chromosomal aberration assay which requires skilled manpower and allows for smaller sample size. The assay can be used for bio monitoring of in vivo genotoxic exposures as well as for *in vitro* experimental exposures to assess the genotoxic potential of a candidate compound and similarly protective effect of a compound that may ameliorate the genotoxicity of a known clastogen. There are a number of confounding factors for the presence and persistence of chromosomal aberrations in the form of micronuclei that should be taken into account for better correlation of extent of actual genetic damage. This is necessary for forming guidelines regarding regulatory requirements as per the WHO and EPA when addressing the question of safety of any chemical agent. Recent modifications in the micronucleus assay techniques have expanded the scope of detecting not only the genetic damage, but also cell proliferation kinetics and differentiation. The practical feasibility of this assay makes it a significant cytogenetic tool.

Chapter 7 - Micronucleus assay (MA) is one of the most potential biomarker for genotoxicity studies. The conventional MA is a highly reliable method for detecting DNA damage in biological cells both *in vivo* and *in vitro*. The history of study of micronucleus (MN) dates back to 19th century. The advantages of MA are immense and hence make it the most followed test in pharmaceutical companies, drug discovery, and regulatory agencies. In the present review, generation of MN in the bone marrow cells of mice induced by various clastogenic chemicals, drugs, and radiation are elaborately elucidated. Bone marrow cells are easily susceptible to oxidative damage and sensitive to various clastogenic as well as aneugenic agents. Majority of the clastogenic agents produces reactive oxygen species (ROS) which attacks and damages the DNA of the bone marrow cells. In mice, the MN is best studied in the early and late maturing stages of RBCs known as the polychromatic erythrocytes (PCEs) and the normochromatic erythrocytes (NCEs). The MN appears in PCEs and NCEs as a tiny acentric fragment of a chromosome or the whole chromosome which lags behind in the cytoplasm during the anaphase of the cell division. Such bodies are easily stained and detected under microscope as small round or oval shaped structure. Sometimes they appear like single or multiple dots taking the same stain as that of the nucleus of the nucleated cells. For MN study two thousand PCEs are scored and the corresponding NCEs are screened as well.

In: Micronucleus Assay: An Overview
Editor: Robert C. Cole

ISBN: 978-1-53616-678-1
© 2020 Nova Science Publishers, Inc.

Chapter 1

CLINICAL APPLICATIONS OF MICRONUCLEUS ASSAY

*Merve Bacanli**

Gülhane Pharmacy Faculty Department of Pharmaceutical Toxicology,
University of Health Sciences, Ankara, Turkey

ABSTRACT

The cells in the human body are exposed to various endogenic and exogenic toxic substances which have the potency to cause genotoxicity. Genotoxicity may result in chronic diseases such as diabetes, cardiovascular diseases, Alzheimer disease, cancer and aging. Because of this, the detection of genotoxicity is very important. Micronucleus (MN) is defined as the small nucleus that forms whenever a chromosome or its fragment is not incorporated into one of the daughter nuclei during cell division. It is concluded that MN assay can be used for risk prediction, screening, diagnosis and treatment of various chronic diseases.

In this chapter, the applications of micronucleus assay will be discussed.

---

* Corresponding Author's Email: mervebacanli@gmail.com, Phone: +90 (312) 304 60 73.

Keywords: genotoxicity, micronucleus, biomonitoring, DNA damage, occupational genotoxicity

1. INTRODUCTION

There are various toxic factors in the environment which have oxidative potential to cause DNA damage in human body cells (Kryston et al. 2011). Every day, human populations are exposed to mutagenic and carcinogenic compounds, both occupationally and/or environmentally (Azqueta et al. 2014).

Genetic biomonitoring of populations exposed to mutagens and/or carcinogens is an important warning system for genetic diseases and cancer. It also allows identification of risk factors. Human genetic biomonitoring studies include various cytogenetic markers including single cell gel electrophoresis (Comet), chromosomal aberration, micronucleus (MN) and sister chromatid exchange (Kassie, Parzefall and Knasmüller 2000).

Micronucleus (MN) is a small nucleus which can be determined under light microscopy (Samanta and Dey 2012). In the last years, MN assay has commonly been used as a biomarker of chromosomal damage, genome instability and cancer risk (Holland et al. 2008).

This chapter aims to provide knowledge about the general information about micronucleus (MN) assay and its applications.

2. MICRONUCLEUS ASSAY

Micronucleus, also known as Howell-Jolly bodies can be seen around the main nucleus within inner half of the cytoplasm at the periphery of the cell. MN was first introduced in 1951 related to acentric fragments (Fenech et al. 2003; Krishna and Hayashi 2000; Kirsch-Volders et al. 2003). This assay has a great statistical power, since it analyzes over 1000 cells (Araldi et al. 2015).

Chromosomal breakage and dysfunction of the mitotic apparatus are the main mechanisms of MN formation (Falck, Catalán, and Norppa 2002). Acentric, autonomously replicating extrachromosomal structures are formed during oncogene amplification in human tumors (Samanta and Dey 2012).

Micronucleus (MN) assay is simple and allows rapid assessment of cells. Cytokinesis-block micronucleus (CBMN) assay uses cytochalasin-B to stop dividing cells from performing cytokinesis (Fenech 1993).

Feulgen, Acridine Orange, May Grunwald Giemsa and Papanicolaou's stain have been used for MN scoring (Thomas et al. 2008).

Exfoliated epithelial cells can be collected easily. These cells are in contact with inhaled and ingested genotoxic agents and carcinogens. Therefore, this cell type is very important in detecting earlier genotoxicity. The genotoxic changes in bronchial, esophageal, cervical, breast and other epithelial cells have been reported in previous studies (Ribeiro et al. 1994).

Buccal mucosa cells are easily accessible tissue for sampling cells in a minimally invasive manner. Buccal mucosa provides a barrier to potential carcinogens. Buccal micronucleus assay has been used to determine the differences between the cytome profiles associated with genetic diseases and ageing (Thomas et al. 2009).

3. APPLICATIONS OF MICRONUCLEUS ASSAY

Exposure to environmental pollutants, radiation, bio-hazard materials, drugs, food/drink habits and free radical injuries may cause MN formation in healthy subjects (Konopacka 2003). Chronic inflammation, chemotherapy, heavy metal poisoning, genetic diseases and nutritional deficiency are also responsible on MN formation (Konopacka 2003; Holland et al. 2007; Lewińska et al. 2007; Burgaz et al. 1999; Thomas et al. 2008; Fenech et al. 2005).

3.1. In Vitro Applications

Li et al. (2012) studied the genotoxic effects of silver nanoparticles (AgNPs) in TK6 human lymphoblastoid cells. At a concentration of 30 µg/ml, AgNPs induced a significant, 3.17-fold increase with a net increase of 1.60% in micronucleus frequency over the vehicle control. In another study, the genotoxic effects of 20 nm titanium dioxide nanoparticles (TiO_2-NPs) were evaluated using Chinese hamster ovary (CHO-K1) cells. It is found that the genotoxicity of TiO_2-NPs is mediated mainly through the generation of oxidative stress in cells (Chen, Yan and Li 2014). Kazimirova et al. (2012) assessed the genotoxicity of poly-lactic-co-glycolic acid–polyethylene oxide copolymer (LGA–PEO) nanoparticles was assessed in TK6 cells. They found no statistically significant differences in the frequencies of micronucleated binucleated cells (MNBNCs) between untreated and treated cells in either of the conditions used. Jomehzadeh et al. (2018) showed that vinblastine and gamma irradiation both were able to significantly increase micronucleated-binucleated cells (MnBi) frequency in L929 cells. In the human lymphocyte and Chinese hamster fibroblast (V79) cells, it is found that galangin, puerarin, ursolic acid, limonene and naringin revealed a reduction in the DNA damage caused by hydrogen peroxide (H_2O_2) (Bacanlı, Başaran and Başaran 2017, 2015).

3.2. Occupational Studies

Torres et al. (2019) evaluated the buccal cell MN frequency after occupational exposure to low doses of ionizing radiation. The frequency of MN increased in exposed group when compared to non-exposed group. MN frequencies of 22 chromate contact workers and 44 non-chromate contact workers from an electroplating enterprise with long-term occupational environment were investigated. It is demonstrated that the average lymphocyte MN frequency of chromate contact workers was higher than the reference value of the general population (Liu et al. 2019). The genotoxic effects of smoking and chewing tobacco studied in the exfoliated cells of tobacco cells. Weak positive and nonsignificant correlation were observed between age and mean percentage of micronucleated cells in smokers and smokers + chewers, respectively, while it was weak negative and nonsignificant in chewers. In control group, correlation between age and percentage of micronucleated cells was weak positive and nonsignificant at 5% level of significance (Upadhyay et al. 2019).

Cakmak et al. (2019) assessed the genetic damage of operating and recovery room personnel occupationally exposed to waste anaesthetic gases in the peripheral blood lymphocytes and buccal epithelial cells. The MN frequencies were increased in both cell types of the healthcare personnel.

The genotoxic effects of cigarette smoke are well known. The MN formation inducing ability of cigarette smoke was demonstrated in Chinese hamster fibroblast (V79) cells. After exposure to smoke containing between 88 and 224 mg/m^3 particulate matter, an induction of MN was measured (Massey et al. 1998).

The DNA damage of Turkish ceramic workers was investigated in the buccal cells. MN frequencies of workers increased compared to their controls (Anlar et al. 2017). Besides, Aksu et al. (2018) showed the increases in the buccal MN frequencies of Turkish welders when compared to their healthy controls.

3.3. Clinical Applications

It is concluded that the prevalence and frequency of spontaneous occurrence of MN increases with the age in the human lymphocytes (Bakou et al. 2002).

Most of the studies have demonstrated that the level of baseline chromosomal damage of untreated cancer patients was significantly higher than healthy controls (Nersesyan 2007). Therefore, MN scoring is important in the screening of high-risk population for a specific cancer (Samanta and Dey 2012).

Micronucleus (MN) assay is a useful tool in virology. The Tax protein of Human T-Leukemia Virus Type-1 (HTLV-I) and II (HTLV-II) induced

MN formation (Semmes et al. 1996; Majone, Semmes and Jeang 1993). Similarly, the MN frequencies of peripheral bloods of human papillomavirus (HPV) infected women are significantly higher (Leal-Garza et al. 2002; Cassel et al. 2014).

The cancer incidence or mortality rates of 6718 subjects from 10 countries screened. The results demonstrated that MN frequency of peripheral blood lymphocytes was a predictive biomarker of cancer risk within a population of healthy subjects (Bonassi et al. 2007). El-Zein et al. (2006) measured the nicotine-derived nitrosamine 4-(methylnitrosamino)-1-(3-pyridyl)-1-butanone (NNK) induced MN frequency. Their results indicated that the CBMN assay was extremely sensitive to NNK-induced genetic damage and might serve as a strong predictor of lung cancer risk. Nikolouzakis et al. (2019) studied the effects of systemic treatment on the MN frequency in the peripheral bloods of metastatic colon cancer patients. The findings of that study suggested that the MN frequency might serve as a promising prognostic/predictive biomarker for the monitoring of the treatment response of colorectal cancer patients.

An increased MN frequency has been demonstrated in peripheral blood lymphocytes of patients with polycystic ovary syndrome, type 2 diabetes mellitus, coronary artery disease and metabolic syndrome (Andreassi et al. 2011). Corbi et al. (2014) investigated the elevated MN frequency in type 2 diabetes, dyslipidemia and periodontitis patients. Buccal MN frequency of chronic and aggressive periondiditis patients were found to be significantly higher than their controls (Zamora-Perez et al. 2015). The increased frequency of MN in chronic obstructive pulmonary disease patients was primarily assigned to clastogenic events and DNA amplification because the frequency of nucleoplasmic bridges and buds was also increased (Maluf et al. 2007). In another study with children who have chronic kidney disease, it was found that centromere negative micronucleus (C− MN) and centromere positive micronucleus (C+ MN) frequencies were significantly higher in each subgroup children (predialysis, regular haemodialysis and transplanted) than in the control group. Additionally, MN frequencies in mononuclear cells, nucleoplasmic bridges and nuclear buds in binucleated cells were increased in children with chronic kidney disease (Cakmak Demircigil et al. 2011).

CONCLUSION

In conclusion, MN is an important biomarker in determining genotoxicity, biomonitoring of important chronic diseases including cancer and genetic diseases. With the advantages of MN assay, it is easy to get large amounts of data in human biomonitoring studies in a very short period. However, it is very important to be aware of its limitations. The future usage of MN assay could impact some other important areas. The MN assay will be a good addition to the currently existing tests for human biomonitoring studies.

REFERENCES

Aksu, İ., Anlar, H. G., Taner, G., Bacanlı, M., İritaş, S., Tutkun, E. and Basaran, N. (2018). Assessment of DNA damage in welders using comet and micronucleus assays. *Mutation Research/Genetic Toxicology and Environmental Mutagenesis*, doi: 10.1016/ j.mrgentox.2018.11.006.

Andreassi, M. G., Barale, R., Iozzo, P. and Picano, E. (2011). "The association of micronucleus frequency with obesity, diabetes and cardiovascular disease". *Mutagenesis,* 26 (1): 77 - 83.

Anlar, H. G., Taner, G., Bacanli, M., Iritas, S., Kurt, T., Tutkun, E., Yilmaz, Ö. H. and Basaran, N. (2017). "Assessment of DNA damage in ceramic workers". *Mutagenesis,* 33 (1): 97 - 104.

Araldi, R. P., Melo, T. C., Mendes, T. B., Sá Júnior, P. L., Nozima, B. H. N., Ito, E. T., Carvalho, R. F., Souza, E. B. and Stocco, R. C. (2015). Using the comet and micronucleus assays for genotoxicity studies: a review. *Biomedicine and pharmacotherapy,* 72: 74 - 82.

Azqueta, A., Slyskova, J., Langie, S. A. S., Gaivão, I. O. and Collins, A. (2014). Comet assay to measure DNA repair: approach and applications. *Frontiers in genetics,* 5: 288.

Bacanlı, M., Başaran, A. A. and Başaran, N. (2015). The antioxidant and antigenotoxic properties of citrus phenolics limonene and naringin. *Food and chemical Toxicology,* 81: 160 - 170.

Bacanlı, M., Başaran, A. A. and Başaran, N. (2017). The antioxidant, cytotoxic, and antigenotoxic effects of galangin, puerarin, and ursolic acid in mammalian cells. *Drug and chemical toxicology,* 40 (3): 256 - 262.

Bakou, K., Stephanou, G., Andrianopoulos, C. and Demopoulos, N. A. (2002). Spontaneous and spindle poison-induced micronuclei and chromosome non-disjunction in cytokinesis-blocked lymphocytes from two age groups of women. *Mutagenesis,* 17 (3): 233 - 239.

Bonassi, S., Znaor, A., Ceppi, M., Lando, C., Chang, W. P., Holland, N., Kirsch-Volders, M., Zeiger, E., Ban, S., Barale, R., Bigatti, M. P., Bolognesi, C., Cebulska-Wasilewska, A., Fabianova, E., Fucic, A., Hagmar, L., Joksic, G., Martelli, A., Migliore, L., Mirkova, E., Scarfi, M. R., Zijno, A., Norppa, H. and Fenech, M. (2007). An increased micronucleus frequency in peripheral blood lymphocytes predicts the risk of cancer in humans. *Carcinogenesis,* 28 (3): 625 - 631.

Burgaz, S., Karahalil, B., Bayrak, P., Taşkın, L., Yavuzaslan, F., Bökesoy, I., Anziom, R. B. M., Bos, R. P. and Platin, N. (1999). Urinary cyclophosphamide excretion and micronuclei frequencies in peripheral lymphocytes and in exfoliated buccal epithelial cells of nurses handling antineoplastics. *Mutation Research/Genetic Toxicology and Environmental Mutagenesis,* 439 (1): 97 - 104.

Cakmak Demircigil, G., Aykanat, B., Fidan, K., Gulleroglu, K., Bayrakci, U. S., Sepici, A., Buyukkaragoz, B., Karakayali, H., Haberal, M., Baskin, E., Buyan, N., Burgaz, S. (2011). Micronucleus frequencies in peripheral blood lymphocytes of children with chronic kidney disease. *Mutagenesis,* 26 (5): 643 - 650.

Cassel, A. P. R., Barcellos, R. P., Silva, C. M. D., Almeida, S. E. M. and Rossetti, M. L. R. (2014). Association between human papillomavirus

(HPV) DNA and micronuclei in normal cervical cytology. *Genetics and molecular biology,* 37 (2): 360 - 363.

Chen, T., Yan, J. and Li, Y. (2014). Genotoxicity of titanium dioxide nanoparticles. *Journal of Food and Drug Analysis,* 22 (1): 95 - 104.

Corbi, S. C. T., Bastos, A. S., Orrico, S. R. P., Secolin, R., Dos Santos, R. A., Takahashi, C. S. and Scarel-Caminaga, R. M. (2014). Elevated micronucleus frequency in patients with type 2 diabetes, dyslipidemia and periodontitis. *Mutagenesis,* 29 (6): 433 - 439.

Çakmak, G., Eraydın, D., Berkkan, A., Yağar, S. and Burgaz, S. (2019). Genetic damage of operating and recovery room personnel occupationally exposed to waste anaesthetic gases. *Human and experimental toxicology,* 38 (1): 3 - 10.

El-Zein, R. A., Schabath, M. B., Etzel, C. J., Lopez, M. S., Franklin, J. D. and Spitz, M. R. (2006). Cytokinesis-Blocked Micronucleus Assay as a Novel Biomarker for Lung Cancer Risk. *Cancer Research,* 66 (12): 6449 - 6456.

Falck, G. C.-M., Catalán, J. and Norppa, H. (2002). Nature of anaphase laggards and micronuclei in female cytokinesis-blocked lymphocytes. *Mutagenesis,* 17 (2): 111 - 117.

Fenech, M. (1993). The cytokinesis-block micronucleus technique and its application to genotoxicity studies in human populations. *Environmental health perspectives,* 101 (suppl. 3): 101 - 107.

Fenech, M., Baghurst, P., Luderer, W., Turner, J., Record, S., Ceppi, M. and Bonassi, S. (2005). Low intake of calcium, folate, nicotinic acid, vitamin E, retinol, β-carotene and high intake of pantothenic acid, biotin and riboflavin are significantly associated with increased genome instability—results from a dietary intake and micronucleus index survey in South Australia. *Carcinogenesis,* 26 (5): 991 - 999.

Fenech, M., Chang, W. P., Kirsch-Volders, M., Holland, N., Bonassi, S. and Zeiger, E. (2003). HUMN project: detailed description of the scoring criteria for the cytokinesis-block micronucleus assay using isolated human lymphocyte cultures. *Mutation Research/Genetic Toxicology and Environmental Mutagenesis,* 534 (1-2): 65 - 75.

Holland, N., Bolognesi, C., Kirsch-Volders, M., Bonassi, S., Zeiger, E., Knasmueller, S. and Fenech, M. (2008). The micronucleus assay in human buccal cells as a tool for biomonitoring DNA damage: the HUMN project perspective on current status and knowledge gaps. *Mutation Research/Reviews in Mutation Research,* 659 (1-2): 93 - 108.

Holland, N., Harmatz, P., Golden, D., Hubbard, A., Wu, Y., Bae, J., Chen, C., Huen, K. and Heyman, M. B. (2007). Cytogenetic damage in blood lymphocytes and exfoliated epithelial cells of children with inflammatory bowel disease. *Pediatric research,* 61 (2): 209.

Jomehzadeh, Z., Haddad, F. and Matin, M. M. (2019). Investigating the Genotoxic Effect of Gamma Irradiation on L929 Cells after Vinblastine Treatment Using Micronucleus Assay on Cytokinesis-blocked Binucleated Cells. *Journal of Cell and Molecular Research,* 10 (2): 52 - 58.

Kassie, F., Parzefall, W. and Knasmüller, S. (2000). Single cell gel electrophoresis assay: a new technique for human biomonitoring studies. *Mutation Research/Reviews in Mutation Research,* 463 (1): 13 - 31.

Kazimirova, A., Magdolenova, Z., Barancokova, M., Staruchova, M., Volkovova, K. and Dusinska, M. (2012). Genotoxicity testing of PLGA–PEO nanoparticles in TK6 cells by the comet assay and the cytokinesis-block micronucleus assay. *Mutation Research/Genetic Toxicology and Environmental Mutagenesis,* 748 (1): 42 - 47.

Kirsch-Volders, M., Sofuni, T., Aardema, M., Albertini, S., Eastmond, D., Fenech, M., Ishidate Jr., M., Kirchner, S., Lorge, E. and Morita, T. (2003). Report from the in vitro micronucleus assay working group. *Mutation Research/Genetic Toxicology and Environmental Mutagenesis,* 540 (2): 153 - 163.

Konopacka, M. (2003). Effect of smoking and aging on micronucleus frequencies in human exfoliated buccal cells. *Neoplasma,* 50 (5): 380 - 382.

Krishna, G. and Hayashi, M. (2000). In vivo rodent micronucleus assay: protocol, conduct and data interpretation. *Mutation*

Research/Fundamental and Molecular Mechanisms of Mutagenesis, 455 (1-2): 155 - 166.

Kryston, T. B., Georgiev, A. B., Pissis, P. and Georgakilas, A. G. (2011). Role of oxidative stress and DNA damage in human carcinogenesis. *Mutation Research/Fundamental and Molecular Mechanisms of Mutagenesis,* 711 (1-2): 193 - 201.

Leal-Garza, C. H., Cerda-Flores, R. M., Leal-Elizondo, E. and Cortés-Gutiérrez, E. I. (2002). Micronuclei in cervical smears and peripheral blood lymphocytes from women with and without cervical uterine cancer. *Mutation Research/Genetic Toxicology and Environmental Mutagenesis,* 515 (1-2): 57 - 62.

Lewińska, D., Palus, J., Stępnik, M., Dziubałtowska, E., Beck, J., Rydzyński, K., Natarajan, A. T. and Nilsson, R. (2007). Micronucleus frequency in peripheral blood lymphocytes and buccal mucosa cells of copper smelter workers, with special regard to arsenic exposure. *International archives of occupational and environmental health,* 80 (5): 371 - 380.

Li, Y., Jian Yan, C., Chen, Y., Mittelstaedt, R. A., Zhang, Y., Biris, A. S., Heflich, R. H. and Chen, T. (2012). Genotoxicity of silver nanoparticles evaluated using the Ames test and in vitro micronucleus assay. *Mutation Research/Genetic Toxicology and Environmental Mutagenesis,* 745 (1): 4 - 10.

Liu, J. X., Hu, G. P., Zhao, L., Zhang, Y. M., Wang, L., Jia, G., Liu, R. X., Feng, H. M. and Xu, H. D. (2019). Early effects of low-level long-term occupational chromate exposure on workers'health. *Journal of Peking University. Health sciences,* 51 (2): 307 - 314.

Majone, F., Semmes, O. J. and Jeang, K.-T. (1993). Induction of micronuclei by HTLV-I Tax: a cellular assay for function. *Virology,* 193 (1): 456 - 459.

Maluf, S. W., Mergener, M., Dalcanale, L., Costa, C. C., Pollo, T., Kayser, M., Basso da Silva, L., Pra, D. and Teixeira, P. J. Z. (2007). DNA damage in peripheral blood of patients with chronic obstructive pulmonary disease (COPD). *Mutation Research/Genetic Toxicology and Environmental Mutagenesis,* 626 (1): 180 - 184.

Massey, E., Aufderheide, M., Koch, W., Lodding, H., Pohlmann, G., Windt, H., Jarck, P. and Knebe, J. W. (1998). Micronucleus induction in V79 cells after direct exposure to whole cigarette smoke. *Mutagenesis,* 13 (2): 145 - 149.

Nersesyan, A. K. (2007). Possible role of the micronucleus assay in diagnostics and secondary prevention of cervix cancer: a minireview. *Cytology and genetics,* 41 (5): 317 - 318.

Nikolouzakis, T. K., Stivaktakis, P. D., Apalaki, P., Kalliantasi, K., Sapsakos, T. M., Spandidos, D. A., Tsatsakis, A., Souglakos, J. and Tsiaoussis, J. (2019). Effect of systemic treatment on the micronuclei frequency in the peripheral blood of patients with metastatic colorectal cancer. *Oncology letters,* 17 (3): 2703 - 2712.

Ribeiro, L. R., Salvadori, D. M. F., Rios, A. C. C., Costa, S. L., Tates, A. D., Törnqvist, M. and Natarajan, A. T. (1994). Biological monitoring of workers occupationally exposed to ethylene oxide. *Mutation Research/Environmental Mutagenesis and Related Subjects,* 313 (1): 81 - 87.

Samanta, S. and Dey, P. (2012). Micronucleus and its applications. *Diagnostic cytopathology,* 40 (1): 84 - 90.

Semmes, O. J., Majone, F., Cantemir, C., Turchetto, L., Hjelle, B. and Jeang, K.-T. (1996). HTLV-I and HTLV-II Tax: differences in induction of micronuclei in cells and transcriptional activation of viral LTRs. *Virology,* 217 (1): 373 - 379.

Thomas, P., Harvey, S., Gruner, T. and Fenech, M. (2008). The buccal cytome and micronucleus frequency is substantially altered in Down's syndrome and normal ageing compared to young healthy controls. *Mutation Research/Fundamental and Molecular Mechanisms of Mutagenesis,* 638 (1-2): 37 - 47.

Thomas, P., Holland, N., Bolognesi, C., Kirsch-Volders, M., Bonassi, S., Zeiger, E., Knasmueller, S. and Fenech, M. (2009). Buccal micronucleus cytome assay. *Nature protocols,* 4 (6): 825.

Torres, A., Armindo dos Santos Rodrigues, L., Linhares, D., Camarinho, R., Páscoa Soares Rego, Z. M. N. and Garcia, P. V. (2019). Buccal epithelial cell micronuclei: Sensitive, non-invasive biomarkers of occupational exposure to low doses of ionizing radiation. *Mutation Research/Genetic Toxicology and Environmental Mutagenesis,* 838: 54 - 58.

Upadhyay, M., Verma, P., Sabharwal, R., Subudhi, S. K., Jatol-Tekade, S., Naphade, V., Choudhury, B. K. and Sahoo, P. D. (2019). Micronuclei in Exfoliated Cells: A Biomarker of Genotoxicity in Tobacco Users. *Nigerian journal of surgery: official publication of the Nigerian Surgical Research Society,* 25 (1): 52 - 59.

Zamora-Perez, A. L., Ortiz-García, Y. M., Lazalde-Ramos, B. P., Guerrero-Velázquez, C., Gómez-Meda, B. C., Ramírez-Aguilar, M. Á. and Zúñiga-González, G. M. (2015). Increased micronuclei and nuclear abnormalities in buccal mucosa and oxidative damage in saliva from patients with chronic and aggressive periodontal diseases. *Journal of periodontal research,* 50 (1): 28 - 36.

In: Micronucleus Assay: An Overview
Editor: Robert C. Cole

ISBN: 978-1-53616-678-1
© 2020 Nova Science Publishers, Inc.

Chapter 2

MICRUNUCLEUS TEST EVOLUTION

Alejandra Hernández-Ceruelos[1,], PhD,*
Sergio Muñoz-Juárez[2], PhD,
Fernando Vázquez-Rivera[1], MD,
Jesús Ruvalcaba-Ledezma[1], PhD
and Luilli López Conteras[1], PhD

[1]Cuerpo Académico de Salud Pública, Medical School,
Instituto de Ciencias de la Salud Universidad Autónoma
del Estado de Hidalgo, México
[2]Departamento de Investigación, Hospital General de Pachuca,
Secretaria de Salud de Hidalgo, México

ABSTRACT

Ionizating radiations, ultraviolet rays, geogenic or anthropogenic pollutants can induce mutagenic, teratogenic or carcinogenic effects due to the induction of micro or macrolession over DNA. Several models have been used to measure the mutagenic and clastogenic effect of such agents; one of them with many advantages is the Miicronucleus test (MN). This assay can be performed *in vitro* and *in vivo*, using vegetal or animal tissue.

---

* Corresponding Author's Email: alejandra.ceruelos@gmail.com.

The test can be performed at a low cost and, in most cases, it does not require highly especialized equipment. The test can be performed under controlled laboratory conditions or as a biomarker of environmental exposure to genotoxic agents. Micronucleus (MN) are defined as intracitoplamatic chromtine fragments separated from the main celular nucleus. The fragments are the result of chromosomal break or chromosomes that had suffered delay during anaphase, then MN has been recognized as biomarker of chromosomic damage. Therfore, the aplication of the MN assay has demosntrated a huge capacity to determine clastogenic or aneugenic effects under different conditions. First studies were performed in mouse bone marrow using histological stains looking for abnormal nucleus in an atempt to replace metaphases analysis. Later the test was performed in mouse pheripherial blood, but MN can be observed also in nucleated cells. In lymphocites cultured with cytocalacine B, as a citokinesis blocker, to obtain binucleated cells can be use as an *in vitro* test to observe micronucleus and other nuclear abnormalities. Another *in vitro* model is the culture of *Tradescantia paludosa* influorescences. Nowadays, the study of exfoliaed ephiteial cell for oral cavity or urinary epithelium has been used for biomonitorization in human population using the cytome model where apotosis induction can also been measure, as well as, binucleated cells and nuclear buds as indicators of chromosomal inestability.

Keywords: micronucleus, cytome, genotoxicity

INTRODUCTION

Micronucleus assay can be performed in diverse *in vitro* and *in vivo* models, such as erytrocites, bone marow cells, lymphocites, vegetal and epitelial cells among others. The guidelines consider the MN test as part of the special trials to determine mutagenicity of possible clastogenic or anegenic agents. In the last years, the test have been used for biomotiring human populations exposed to agrochemical and polutants, and also as a biomarker of genetic hazard in special populations.

Another advantage of the test under controlled experimental conditions is the capacity to determine the mutagenic potential in accute or chronic exposition, depending on the design of the project and the kind of chemical substance to be tested. It is advisable the use of a negative control, a positive control and at least 3 doses to test, being the highest one the equivalent to ¼ of the lethal dose 50 in animals. For the design of the protocol for acute, subchronic or chronic exposures of the genotoxicant, it is necessary to consider the natural rout of intoxication, the source of exposition and the time that in normal conditions the agent would be in contact with the subjects. For *in vitro* test, it is advisable to test the chemical agent with and without bioactivation with S9 fraction. For *in vivo* test, the use of five to ten animals per group is neccesary to perform ANOVA and mean differences as the sugested statistical analysis.

For complex mixtures or environmental exposition it is importat to determine the impact of the life style, gender, nutritional state, unhealty habits, chronic deseases such as Diabetes mellitus, among other factors to minimize the bias that could lead to confusing data. Therefore, for human population it is important to survey the participants and the comparison with a control population is always needed, and at minimum of 30 subjects per group in order to be accurate in the statistic test.

The flexibility of the model has made the MN test one of the most useful models in genetic toxicology. In this chapter, we present the fundaments, historical development and current applications of the assay.

MICRONUCLEUS MECHANISM OF FORMATION

Micronucleus (MN) is an abnormal structure in eukaryotic cells. It is originated from chromosome fragments or whole chromosomes that lag behind at anaphase during nuclear division under physical and chemical factors. The MN index in rodent and human cells has become one of the standard cytogenetic endpoints and biomarkers used in genetic toxicology *in vivo* or *in vitro*. (Gang et. al. 2018).

MN are cytosolic chromatin structures that are compartmentalized by a nuclear envelope. Micronuclei are originated during mitosis, either because whole chromosomes separate aberrantly, or because DNA damage generates acentric chromosomal fragments and thus are unable to migrate on ahaphase. The resulting structures rebuild their own nuclear envelope away from the main chromatin mass. Micronuclei can persist for multiple cell divisions, but the ennvelope can be distroyed during G2 phase. A MN can be reincorporated into the primary nucleus, creating the conditions for DNA fragments to rejoin the main genome at random locations, this phenomena is called chromothripsis. Therefore, an intact envelope around a micronucleus maintains the integrity of its genetic material and thereby protects against chromothripsis (Kiraly et al., 2016). Chromothripsis has been related to the initiation and development of human cancer (Hovhannisyan, et al., 2018).

To explain the origin of micronucleus at molecular level, we have to consider this mechanims: DNA lesions leading to DNA double strand breaks, DNA replication stress or telomere erosion may result in chromatin bridges or acentric fragments at anaphase that will form MN on mitotic exit. On the other hand, different missegregation events may explain the origin of whole-chromosome carrying MN, including merotelic kinetochore-microtubule interactions, kinetochore damage, spindle disruption and cytokinesis inhibition. All of them lead to micronucleus formation (Fenech et al., 2011; Kisurina-Evgenieva et al., 2016).

Micronuclei containing whole chromosomes/chromatids may be formed after failure of the mitotic spindle, leading to misattachments of microtubules on kinetochores or other parts of the mitotic apparatus. Erroneous kinetochore attachments such as monotelic and merotelic kinetochore orientations can lead to MN formation. Monotelic orientation occurs when only one sister kinetochore attaches to the spindle microtubules. Merotelic kinetochore attachments are defined by the persistent attachment of microtubules from both spindle poles to a single chromosome, if the number of attached microtubules from both poles is nearly equivalent, the chromatid will not move in either direction at anaphase, producing a lagging chromatid, yielding a micronucleus (Kirsch-Volders et al., 2011).

Micronuclei harbouring chromosomal fragments can arise by clastogens that directly induce double-strand DNA breakage (DSB), conversion of single-strand breaks (SSB) into DSB after cell replication, or inhibition of DNA synthesis. Furthermore micronuclei may have their origin in fragments derived from broken anaphase bridges formed due to chromosome rearrangements such as dicentric chromatids, intermingled ring chromosomes or union of sister chromatids (Norppa and Falck 2003).

Once generated as whole chromosomes or chromosome fragments lagging behind at anaphase, MN can encounter different destinies, including reincorporation into the main nucleus or they can suffer extrusion from the cell, interphase degradation or persistence along several cell generations. The relative frequencies of these different fates and the underlying mechanisms have been the subject of numerous studies starting already in the 1990's. These early studies showed that MN can perform DNA synthesis and can be reincorporated into one of the nuclei at the following mitosis, giving rise to further abnormal karyotypes. Starting from around 2010, the widespread use of time lapse microscopy has provided new important insights into the fate of MN and their potential role in chromosome instability (Russo and Degrassi, 2018).

Zhang et al., (2015) demonstrated a direct association between MN and chromothripsis by combining live-cell imaging and whole-genome sequencing (a combined technique referred to as 'LookSeq'). The authors induced micronucleus formation by transiently treating cells with nocodazole, a chemical agent that destabilizes the mitotic spindle and increases the frequency of improper chromosome attachment, lagging chromosomes and MN. The authors observed micronucleated cells that underwent micronuclear rupture during S phase and that subsequently divided to produce two daughter cells. But how could the researchers determine whether being in a micronucleus affects chromosomal structure? They showed that DNA in MN was not replicated properly and thus assumed that the micronuclear chromosome would not be equally distributed to the two daughter cells after cell division. This process would generate an asymmetry in the number of chromosomes (copy number) and their parental origin (haplotype).

Chromothripsis is a recently described phenomenon in which multiple genomic rearrangements are generated in a single catastrophic event. These MN undergo defective and asynchronous DNA replication, resulting in DNA damage and often extensive fragmentation of the chromosome. Pulverization of chromosomes in MN may be one of explanations for chromothripsis in cancer and developmental disorders. Isolated in MN chromosomes undergo massive local DNA breakage and rearrangement. Chromosomes within MN can be reincorporated into daughter nuclei following mitosis and the remaining MN persisted in cells well into the second generation. Thus, mutations, being present in MN, can be reincorporated into a stably genome (Hovhannisyan, et al., 2018).

As consequence of this damage, MN are considered as early biological predictors of carcinogénesis, since whole-chromosome aneuploidy is a major feature of cancer genomes, yet its role in tumour development remains controversial. This contrasts with chromosome breaks and rearrangements, which are known to produce cáncer by causing mutations. Recent genetic evidence demonstrates that increased rates of whole-chromosomemis-segregation can accelerate oncogénesis; however, the only established mechanism by which whole-chromosome segregation errors promote tumorigenesis is by facilitating the loss of heterozygosity for tumour suppressors (Crasta et al., 2012).

It is believed that when the process of equal division of genetic material after reduplication is disturbed. The distribution in either of the two nuclear parts is not distributed into a new nucleus, but may form a smaller micronucleus and a larger macronucleus. To contradict this view the protozoans (amoebas, ciliates, flagellates, sporozoans) have often two types of nuclei with distinct functions. In paramecia the macronucleus is the centre of all metabolic activities of the organism. Comparative analysis of the macronucleus and micronucleus of Tetrahymena has yielded histone variants, and other properties of chromatin now recognized to be general features of the eukaryotic biology (Kiraly et al., 2016).

MN in different mammalian cell lines as normal intermediates of chromatin condensation. Chromatin condensation starts with the supercoiling of the highly decondensed chromatin structrure referred to as chromatin veil, turning around itself and emerging first as a chromatin plate, which is then extruded as a larger supercoiled micronucleus. In early S phase the micronucleus appears as the head portion of the more condensed chromatin ribbon, which continues in the two arms of the continuous chromatin structure. The turned around head portion is folded back to a linear structure at the end of the S phase when the U, V and 8 shaped (8 is open at the bottom) elongated chromosomes are becoming long linear prechromosomes. After the head portion (MN) is folded back and the linear ribboned chromatin is formed in late S phase, the micronucleus is not visible anymore. (Kiraly, et al., 2016).

The potential postmitotic fates of MN after their formation in the micronucleated cell is poorly understood and possibilities and according to Kirsch-Volders et al., (2011) they migth include:

a) retention within the cell's cytoplasm as an extra-nuclear entity, when MN may complete one or more rounds of DNA/chromosome replication
b) re-incorporation into the main nucleus (when the reincorporated chromosome may be indistinguishable from those of the main nucleus and might resume normal biological activity).
c) elimination of the micronucleated cell as a consequence of apoptosis.
d) expulsion from the cell (when the DNA within the MN is not expected to be functional or capable of replication owing to the absence of the necessary cytoplasmic components).

MICRONUCLEUS AS GENOTOXIC MODEL

Several guidance documents on genotoxicity testing strategies have been issued by advisory bodies and regulatory agencies for pharmaceuticals via the International Conference on Harmonisation (ICH). To be valid, it is needed at least two *in vitro* tests covering bacterial gene mutation (Ames test) and chromosome damage (*in vitro* chromosomal aberration (CA) or micronucleus (MN) test) and one or more *in vivo* tests (e.g., *in vivo* MN test, the transgenic rodent mutation (TGR) assay, *in vivo* comet assay) will be required in the tired approach or standard test battery system for identification of the genotoxic substances in the guidance, especially in case of *in vitro* positive results (Morita et al., 2016).

The micronucleus assay is recognized as a reliable method for the screening of chemical carcinogens. An increase in the frequency of micronucleated cells is an indication of induced chromosome damage and apoptosis. The three most important clinical biomarkers for DNA damage are micronuclei, DNA strand breaks using the comet assay, and the base modification of 8-oxodG. The only reliable marker with an established potential to predict disease complications is micronucleus frequency. MN assay is a well standardized approach for evaluation of mutagens and used to characterize the chromosomal content of micronuclei (Schupp et al., 2016).

IN VIVO MICRONUCLEUS TEST

The MN was recognized in the end of the 19th century when Howell and Jolly found small inclusions in the blood taken from cats and rats. The small inclusions, called Howell-Jolly body, were also observed in the erythrocytes of peripheral blood from severe anemia patients (Hayashi, 2016).

In 1959, Evans et al., reported that gamma-rays induced micronuclei in root tips of kidney beans, and tried to evaluate the chromosomal aberration quantitatively. This was the first report to evaluate chromosomal aberration by the frequency of cells harboring micronucleus among normal cells and they estimated that about 60% of the chromosomal fragments contributed to micronucleus formation.

In 1970, Boller and Schmid developed a test method to evaluate the frequency of micronucleated erythrocytes among normal erythrocytes, which lack their own nuclei during hematopoiesis, using bone marrow and peripheral blood cells of Chinese Hamster treated with a strong alkylating agent, trenimon. In the paper, they named this method as "Mikrokern-Test (micronucleus test)." Heddle's et al., (1973) and Schmid et al., (1975) set up the basis for the *in vivo* micronucleus test to measure chromosomal damage in bone marrow erythrocytes.

In 1979 two teams, Cole et al., and King and Wild observed induction of micronucleus which appears in the fetal mouse liver and peripheral blood cells at the very late stage of gestation whose mothers were treated intraperitoneally with a clastogenic chemical. Some chemical agents are metabolized in the liver and become their active form. When the active forms are unstable and diminish before reaching the bone marrow, the clastogenicity of these chemicals will not be detected by the usual method. Using fetal micronucleus method, however, the unstable active metabolites (e.g., dialkyl-nitrosamine) could be detected.

Micronucleus are hardly seen in the peripheral blood of rats and humans because erythrocytes including micronuclei are captured and destroyed by the spleen rapidly and effectively. In mice, however, micronucleated erythrocytes exist just the same as normal cells in the peripheral blood. The assay using bone marrow evaluates an acute effect of chemicals but the method using mouse peripheral blood erythrocytes can evaluate a chronic effect of the test chemical by analyzing of mature erythrocytes which harboring micronuclei up to their life span. (Hayashi, 2016). Later in 1983 MacGregor et. al. proposed the circulating erythrocytes mouse model for the micronucleus test as routine toxicity test. This model offered the possibility of measuring the MN frequency for a longer period of time, allowing acute and subacute treatments to determine the genotoxicity and cytotoxic capacity of different agents, and also the use of fluorescent stain using Hoechst 33258 and pyronin Y to identify not only MN but also immature erythrocytes. On the same year Hayashi et al., (1983) reported a fluorescent staining method using acridine orange to identify specifically micronuclei by yellowish green fluorescence emitted from DNA concomitantly to identify immature erythrocytes by red fluorescence emitted from RNA.

As early as 1986, the automation for the analysis of micronucleated erythrocytes has been approached by flow cytometry (Hutter and Stöhr, 1982; Hayashi et al., 1992; Denrtinger et al., 2011) and by image analyzer systems (Romagna and Staniforth, 1989; Asano et al., 1998). Using the flow cytometry, the dose-response relation was studied with ionizing radiation until very low dose level (Abramsson-Zetterberg et al., 1999). These kind studies can be practically because the flow cytometer can analyze a large number of cells in a short period, accordingly the statistical power to detect micronucleus induction would increase. Dertinger and his group is most widely used among automating scoring systems. Their system includes gating for young erythrocyte by immunostaining of erythrocyte membrane and also gating the amount of DNA. Usually, they analyze 20000 young erythrocytes, but a million cells can be analyzed without any difficulty. This means the statistical power can be increased easily when the cell is considered as a unit of analysis (Dertinger et al., 2007).

Erythrocyte MN assay was the development of an efficient staining method to identify newly formed erythrocytes [polychromatic erythrocyte (PCE)] unequivocally. Giemsa staining has been used successfully for this assay and with this stain, PCEs were distinguished from normochromatic (mature) erythrocytes by the colour, i.e., PCEs showed bluish color and normochromatic ones showed pinkish. The classification of cells, however, required a subjective decision by the investigator and that was sometimes difficult to make. Immature erythrocytes contain RNA in their cytoplasm and can be distinguished easily from mature erythrocytes, which do not fluoresce because they lack RNA. The MN is the only element that contains DNA in the mammalian erythrocyte and it can therefore be identified clearly and specifically. This DNA-specific staining has overcome the problem of distinguishing true MN from artefactual MN-like particles and is now recommended for use in worldwide regulatory guidelines (Heddle et al., 2011).

The *in vivo* micronucleus assay, standardized by the OECD guideline No. 474 (2006), where it is described as widespread used genotoxicity test for the detection of clastogenic and aneugenic chemicals. The protocol, conduct and data interpretation were described in detail by Krishna and Hayashi (2000). From a biostatistical perspective the design is a randomized one-way layout, including three or more doses of the test substance, a negative control and a positive control, where five to ten animals were randomized to each group. Commonly, the both sexes are analyzed independently. The primary endpoint is the number micronucleited erythrocytes (MN) per a certain number of scored polychromatic erythrocytes (PCE), per animal (Hothorn and Gerhard, 2009).

For assessing cytogenetic damage, FCM-scoring of micronucleated erythrocytes in rodents is now a well-established and efficient platform which is highly predictive towards carcinogenicity, relatively easy to evaluate and contains improved technical advantages (Dertinger et al., 2011). The strength of this test is that it exclusively detects MN arising in the bone marrow and thus indicates that the genotoxic substance tested is effective in that tissue. Because the hematopoietic cells undergo rapid division, the test is highly sensitive to genotoxic agents as well as to aneugenic agents that produce changes in the chromosome number (Morita et al., 2016).

TRADESCANTIA MICRONUCLEUS ASSAY

The genus *Tradescantia* belongs to the Commelinaceae family and comprises 500 species, which are found mainly in subtropical and tropical areas (Watson and Dallwitz, 1994). The most frequently used clone for genotoxicity studies is #4430, which is a hybrid between *Tradescantia hirsutiflora* and *Tradescantia subacaulis* and was established in the 1960s. The different cultivars of *Tradescantia ssp.* which are suitable for MN experiments can be grown outdoor in areas with moderate climate while in countries with cold winters and snowfalls, they have to be cultivated in the green house (Ma et al., 1994). In this context, it is notable that it has been shown that low (11°C) as well as high (42°C) temperatures may have an impact on the background frequencies and can affect also the results of experiments with genotoxins (Klump et al., 2004).

The Tradescantaia MN Assay (Trad-MCN) is currently a widely used plant bioassay for environmental biomonitoring. This bioassay was originally developed as a test system for the gaseous mutagen 1,2-dibromoethane (Ma et al., 1978). The major advantage of this test procedure in comparison to cytogenetic tests with mammalian cells and bacterial mutagenicity assays. for water testing is, that the plants can be exposed directly to the natural state of the water sample without any concentration procedure or filtration process in the laboratory (Steinkellner et al., 1998).

It is based on the formation of micronuclei resulting from chro- mosome breakage in the meiotic pollen mother cells of *Tradescantia* ssp.

inflorescences (Klumpp et al., 2004). To find out if Trad MN assays can be used for environmental monitoring of soils and sediments which are contaminated with radionuclides and for the detection of polluted artifacts several series of experiments were performed with samples from different regions (India, Brazil, Czech Republic); furthermore, we included also human artifact (tableware) which contained radioisotopes. The plants were exposed under different conditions, i.e., indirectly or via direct contact with the material (Misík et. al 2016).

The inflorescences are fixed in aceto-alcohol (1:3) after a 24 or 30 hr recovery time. After 24 hr of fixation, the samples are stored in 70% ethanol. Since the MCN to be scored are in the wellsynchronized early tetrad stage, therefore, only the buds which contain early stage tetrads (four cells encased in an envelope) are selected for preparation of microslides. The selection of proper buds is done on the basis of size and relative position among the series of buds in a given inflorescence. A pair of dissecting needles are used to remove the glumes and expose the anthers. The meiotic PMC are released from anthers by crushing the anthers on the slide. A drop of aceto-carmine stain is applied over the content of the crushed anthers and allow 2-5 min to stain the nuclei of the cells. When the correct stage of PMC (tetrad stage) is secured by the help of a magnifying lens or a low power dissecting microscope, the debris from the broken anther wall or stamen hair are carefully removed with the dissecting needles before application of the cover glass. The slide with the cell contents under the cover glass is heated (below boiling) repeatedly over an alcohol flame to improve the staining quality. Gentle pressure is applied over the coverglass with the palm of hand which is protected from excessive heat by several layers of absorbent paper. This temporary slide is ready to be scored under 400 x magnification (Ma et al., 1981).

According to the standard protocol, at least 300 tetrads should be evaluated from each inflorescence and it was recommended to employ analysis of variance and Dunnett's T test for the statistical evaluation (Misík, et al., 2011). Other protocols take 20 flower buds with tetrad state pollen, considering 10 slides per treatment group. Anova a Tukey test are suggested for the statistical analysis (Resende de Morais, et al., 2019).

IN VITRO MICRONUCLEUS TEST

The relationship between genome stability and human health becomes most obvious in diseases typically characterized by progressive deterioration of specific tissues, susceptibility to cancer, chromosomal rearrangements, and hypersensitivity to genotoxic agents. The identification of diseases that might cause genetic changes, including chromosomal instability is an important diagnostic criterion that contributes to a better understanding of disease etiologies and the choice of treatment (Hovhannisyan et al., 2018).

There is now an extensive amount of data to support the validation of the *in vitro* micronucleus assay using various cell lines or human lymphocytes. These include, in particular, the international validation studies co-ordinated by the French Society of Genetic Toxicology (SFGT) and the report of the *in vitro* micronucleus assay working group. Development of the cytokinesis-block methodology, by addition of the actin polymerisation inhibitor cytochalasin B during the targeted mitosis, allows the identification of nuclei that have undergone one division as binucleates. This allows the study of mechanisms of micronucleus induction by a combination of the cytokinesis-block method with immunochemical labelling of kinetochores, or hybridization with general or chromosome specific centromeric/telomeric probes (OECD Guidelines, 2006).

Micronucleus assay in human peripheral lymphocytes is usually used to assess chromosomal damage caused by exposure to different environmental, occupational or lifestyle factors and for *in vitro* genotoxicity testing (Patino-Garcia et al., 2006). The use of *in vitro* cell cultures for genotoxic and cytotoxic evaluation is rather economic and they are highly sensitive methods for the early detection of chemical exposure and toxicity. The cytokinesis-block micronucleus (CBMN) assay in human lymphocytes has become one of the most widely used methods for measuring structural and numerical chromosomal changes in human cells *in vitro* and *in vivo*. The use of the CBMN assay in *in vitro* genetic toxicology testing is well established and in fact it has become an accepted standard method to assess the genotoxic hazard of chemicals which led to the development of an OECD (Organisation for Economic Cooperation and Development) guideline for this purpose (Heddle et al., 2011; OECD Guidelines, 2006; Fenech et al., 2011).

The conventional use of the CBMN assay in human biomonitoring of chemical genotoxin exposure involves the collection of a blood sample following acute or chronic *in vivo* exposure to the suspected chemical agent or complex mixture. The lymphocytes in the blood sample are then stimulated to divide *ex vivo* using a mitogen, and MN scored in cells that have divided once which are recognised by their binucleated appearance after cytokinesis-block (using cytochalasin-B) of the cells from the first mitotic cycle (Kirsch-Volders, et al., 2016).

It is known that greater proportion of lymphocytes respond to mitogen in cultures of lymphocytes from young subjects than in those from elderly individuals. Then, MN frequency depended on the proportion of lymphocytes that responded to mitogen as well as the number of divisions that occurred during the culture period prior to harvesting cells. Cytochalasins acts as inhibitors of cytokinesis by blocking polymerisation of actin into the microfilament ring required for binucleate cell fission, and it could inhibit cytokinesis most efficiently in lymphocytes and other mammalian cells (Heddle et al., 2011). Restricting scoring of micronuclei in binucleated cells prevents confounding effects caused by suboptimal cell division kinetics, which is a major variable in this assay (Fenech, 2007).

Many studies showed increased MN frequency in un treated patients with cancer, neurodegenerative diseases, cardiovascular disease, diabetes. Moreover, elevated MN frequencies in peripheral lymphocytes of healthy subjects have been shown to reflect genomic instability and a higher risk of developing cancer later in life, suggesting a predictive role of the assay (Bolognesi et al., 2015). As far as human chromosomes are concerned the most prevalent in micronuclei are chromosomes X, 9, 1 and 16 due to the heterochromatic breakage sites in these chromosomes (Kiraly et al., 2016).

The inhibition of cytokinesis by cytochalasin B allows one to discriminate between cells that did not divide after treatment and cells that did divide, thus preventing the confounding effects caused by differences in cell division kinetics. Because cells are blocked in the binucleated stage, it is also possible to measure nucleoplasmic bridges originating from asymmetrical chromosome rearrangements and/or telomere end fusions as well as nuclear buds that represent a mechanism by which cells remove amplified DNA and that is therefore considered a marker of possible gene amplification (Randa et al., 2011).

The following cell lines may be used; CHL/IU, CHO, SHE and V79 (16, 22, 24, 31). Also mouse lymphoma L5178Y cells may be used although it should be noted that there is some concern about the possible interactions when using cytochalasin B. The use of other cell types should be justified. Since the background frequency of micronuclei will influence the sensitivity of the assay, it is recommended that cell types with low and stable background frequency of micronuclei (lower than 25/1000 cells without cytochalasin B and 30/1000 cells with cytochalasin B) are used in these studies. When human peripheral lymphocytes are used, blood from two different healthy, young (less than 35 years of age), non-smoking donors should be used where possible; blood should not be pooled. In any repeat experiment the same two donors should be used. The background frequency of micronuclei should be within the historic negative control range for the laboratory. Cells should be exposed to the test substance both in the presence and absence of an appropriate metabolic activation system. The most commonly used system is a co-factor-supplemented postmitochondrial fraction (S9) (OECD Guidelines, 2006).

Spontaneous or baseline MN frequencies in cultured human lymphocytes provide an index of the accumulated genetic damage that has occurred during the lifespan of circulating lymphocytes. For the purpose of biological dosimetry, the spontaneous MN frequency refers to the incidence of MN observed in the absence of the environmental risk or exposure that is being assessed (Fenech, 1993).

The CBMN assay has already been proven to be an effective tool for the study of cellular and nuclear dysfunction caused by invitro or invivo ageing, micronutrient deficiency or excess, genotoxin exposure and genetic defects in genome maintenance. More recent studies indicate that this method is likely to also prove fruitful in the emerging fields of nutrigenomics and toxicogenomics and their combinations as it becomes increasingly clear that nutrient status also impacts on sensitivity to exogenous genotoxins (Fenech, 2007).

Moreover, it should be noted that performing the MN assay with cytokinesis-block provides the possibility of measuring important complementary genomic instability events such nucleoplasmic bridge formation, a biomarker of DNA break misrepair or telomere end fusion, which can only be measured in binucleated cells (Fenech, 2007) and to distinguish between background genetic damage (in mononucleated cells) and damage induced during *in vitro* mitosis (in binucleated cells) (Kirsch-Volders et al., 2011).

CBMN as a "cytome" assay is able to determine chromosomal instability, since it is related to the concept implication that every cell in the system studied is scored cytologically for its viability status (necrosis, apoptosis), its mitotic status and its chromosomal instability or damage status, like the presence of micronuclei, NPBs, nuclear buds and the number of centromere probe signals amongst nuclei of binucleated cells if such molecular tools are used in combination with the assay it is used to measure chromosomal instability phenotype, BFB cycles, DNA misrepair, chromosome breaks and asymmetrical rearrangement, telomere end-fusions, cell cycle checkpoint malfunction, malsegregation of chromosomes (when used in combination with pancentromeric or chromosome-specific probes), gene amplification and nuclear elimination of excess DNA, DNA hypomethylation (if combined with centromeric probes for chromosomes 1, 9 and 16), altered mitotic activity and/or cytostasis and also cell death by necrosis or apoptosis (Fenech, 2007).

HUMN PROJECT

The human micronucleus (HUMN) project (http://www.humn.org), established in 1997, is an international collaborative program aimed at studying the micronucleus (MN) frequency in human populations, and assessing the effects of protocol and scoring criteria on the values obtained. The initial focus of the project was the analysis of MN in peripheral lymphocytes from unexposed and exposed individuals, primarily because this was a well-established human test system at the time the project began. These objectives have been achieved for lymphocytes by providing a detailed description of the scoring criteria and by assessing sources of variability for the cytokinesis-block MN assay through a validation effort undertaken by 34 laboratories from 21 countries (Holland et al., 2008).

In 2007 the international HUMN (Human Micronucleus) Project coordinating group launched the HUMNxl ('XL' designating eXfoLiated cell) project to validate the micronucleus (MN) assay specifically in exfoliated buccal cells following the same strategy which had been used earlier for the validation and standardization of the cytokinesis-block micronucleus. In the last 15–20 years the MN assay has been applied to evaluate chromosomal damage for biological monitoring of human populations exposed to a variety of mutagenic and carcinogenic chemical or physical agents. A broad range of baseline MN frequencies has been reported (0.05–11.5 MN/1000 cells) with the majority of values between 0.5 and 2.5 MN/1000 cells. There is no clear pattern of the variations among laboratories from different countries. Many studies report a statistically significant elevation of MN levels in exposed individuals compared to control groups, although the observed effects are relatively small, ranging between 1.1- and 4-fold (Holland, 2008).

In 2009, a detailed standardized protocol for buccal cell collection, slide preparation and scoring was established by the HUMNxl project consortium by taking into account the available procedures, confounding factors and staining artifacts. Additional scoring of all of the cell types and other nuclear anomalies were incorporated into the assay which thus evolved into its current form as a "cytome" assay of DNA damage, cell proliferation, differentiation and cell death (Bolognesi 2013).

BUCAL MICRONUCLEUS CYTOME

The use of surrogate cells, other than lymphocytes, such as exfoliated cells from epithelial tissues is of particular interest because they can be collected with non-invasive methods and is being explored with the aim to evaluate their suitability in biomonitoring studies (Holland et al., 2008: Nersesyan et al., 2014). The application of the MN test in uncultured buccal exfoliated cells, started in the 1980s, to assess local exposure to genotoxic agents, impact of nutrition and lifestyle factors. Increased MN frequency was detected in buccal cells of subjects affected by cancer-associated congenital syndromes characterized by defects in genes encoding for DNA-repair processes, such as ataxia telangiectasia (Rosin and Ochs, 1986; Rosin et al., 1989) suggesting a role of the assay in detecting chromosomal instability. Nowadays, the test is widely applied in biomonitoring inhalation and local exposure to environmental and occupational genotoxic agents. The oral epithelium is the target for the development of lesions characterized by different clinical outcome and grade of malignancy due to the frequent exposure to mechanical, chemical and thermal insults (Lee et al., 2006; Saran et al., (2007). Some of oral lesions such as leukoplakia, oral lichen planus and oral submucous fibrosis were defined as premalignant and were associated with an increased risk of developing oral squamous cell carcinoma (Bolognesi et al., 2015).

Buccal cells constitute the first point of contact for the inhalation or ingestion route and are capable of metabolizing carcinogens to reactive products. MN in buccal cells are a valid method for the detection of cancer risk in humans, as many tumors derive from epithelial. With respect to oral mucosa, it has been described that MN are observed in the basal layer of epithelial tissues, and their presence can reflect genotoxic damage that occurred up to 14 days prior to sample collection Moreover, buccal cells have been shown to have limited DNA repair capacity and therefore may more precisely reflect genomic instability (Hovhannisyan et al., 2018).

Exfoliated buccal mucosa cells can be collected using a wooden tongue-depressor, a metal spatula, or a cytobrush moistened with water or buffer to swab or gently scrape the mucosa of the inner lining of one or both cheeks. Cytobrushes appear to be most effective for collecting large numbers of buccal cells observed that MN frequencies were higher when cells were collected by vigorous, rather than by light, scraping, suggesting a decreasing

MN frequency gradient from basal to superficial layers of mucosa (Holland et al., 2008).

Higher MN frequencies were observed in studies using Giemsa or aceto-orcein staining confirming that the low specificity of these stains leads to false positive results when compared to DNA specific stains such as the most commonly used Feulgen method. For these reasons it was recommended that the Feulgen nuclear staining method together with Fast Green for cytoplasm staining is used as the standard method. The large majority of available studies on the use of MN assay in buccal cells, including some recent ones, focus only on the MNi frequency, however, it has become evident that the comprehensive cytome approach, although more laborious, provides richer and better diagnostic information for identifying and distinguishing between genotoxic, cytostatic and cytotoxic effects but its implementation requires clear and easy to use guidelines to identify and measure all the cell types and nuclear anomalies (Bolognesi et al., 2013).

In humans, MN can be easily assessed in exfoliated epithelial cells (e.g., oral, urothelial, nasal) to obtain a measure of genome damage induced *in vivo,* since this type of exfoliated cells are originated from rapidly dividing epithelial tissue without the need for *ex vivo* nuclear division, so that the cell cultures required for cytogenetic assays based on analysis of metaphase chromosomes, such as chromosome aberrations and sister chromatid exchanges, are not needed (Holland, et al., 2008). Also the sampling techniques to obtain epithelial cells do not imply invasive procedures as mentioned before.

Stem cells in the stratum basale (also called *Stratum germinativum*) produce the basal cells that mature into transitional differentiated cells and then into terminally differentiated cells migrating to the superficial layer after 7 to21 days where the nucleus may eventually disintegrate completely leading to karyolitic cell formation. The basal, transitional and differentiated cells show nuclear alterations associated with different biological processes. They include biomarkers of DNA damage (MNi and nuclear buds); cell death (condensed chromatin, karyorrhexis, pyknotic and karyolytic cells); cytokinetic defects or arrest (binucleated cells) and proliferative activity (basal cell frequency). The cell death nuclear morphologies are not usually seen in basal cells (Bolognesi et al., 2013).

As biomarkers of DNA damage, besides MN, in this model nuclear buds (NBUD) are related to genotoxicity and originate from processes similar to those that give rise to MNs; in addition, NBUDs are involved in the processes of chromosomal DNA break and are formed as a consequence of gene amplification or excess DNA that has been extruded from the nucleus, which can also generate MN. The mechanism of binucleated (BN) cells formation is not clear, but it is likely occurs due to a failure in cytokinesis during cell division (Gómez-Meda et al., 2017). and nucleoplasmic bridges (NBP) reflect formation of dicentric chromosomes (Fenech, 2007) The baseline frequency of MN and other nuclear abnormalities in buccal cells from healthy subjects and the role of technical and biological confounding factors relevant for its variability need to be clearly defined in order to improve the sensitivity and potential specificity of the assay. The biological meaning of other nuclear alterations as biomarkers of different toxic or genotoxic events or as predictive parameters for cancer or other degenerative diseases needs to be further explored (Bolognesi, 2013).

To determine cellular dead, the different stages of apoptosis can be determine in this model as follows (Bolognesi et al., 2013; Gómez-Meda et al., 2017):

- Normal cells (NC): they are differentiated cells, with hexagonal shape and the nucleus look like 1/3 of cellular surface, stained in pink with fluffy appearance.
- Condensed Chromatin (CC): Nucleus is more compact slightly smaller and stained fuchsia color.
- Karyorrhectic cells (KX): is characterized by chromatin aggregation that appears as a mottled nuclear pattern (purple or fuchsia and green), and it is associated with nuclear fragmentation followed by nuclear disintegration.
- Pyknotic nucleus (Ns): are observed as dense nuclear material, normally in deep purple color and very small size and but the mechanism of formation is unknown.

- Kariolysis (KL): is represented by the absence of cellular nuclei and is associated with cell death, and they are also called ghost cell.

The number of buccal mucosa cells to be scored in order to obtain statistically significant results needs to be addressed. The first studies from the 1980s evaluated a relatively low number of cells (~500). Most published studies have scored between 1000 and 3000 cells, although it has been suggested that 10,000 cells may be needed to observe a statistically significant, 50% increase, in MN frequency (Holland et al., 2008). In human populations it is necessary to realize a comparative analysis among a control population and a exposed one. It is suggested for bivariate analysis to use ANOVA, Student T test or Mann-Whitney U test for mean differences.

A systematic and adequately powered investigation of key variables such as age, gender, genotype, season, diet, oral hygiene (e.g., hydrogen peroxide in tooth paste) and dental health (e.g., high number of missing teeth, periodontal status, etc.), lifestyle, smoking, alcohol, and other recreational drugs, needs to be performed to identify the variables that have to be controlled (Holland et al., 2008). For these confusing and intervening variables, a multivariate linear regression analysis is advisable, to determine their possible effects on the results.

A methanalysis review performed by Bolognesi et al., (2015) reported the clinical applications of the MN test in exfoliated buccal mucosa cells in patients with oral, head-and-neck, breast, bladder, and other cancers, oral premalignant or non-malignant diseases, diabetes, different chronic diseases. Increase of MN number in buccal cells of individuals using various tobacco forms (smokeless tobacco and smoking tobacco) with potentially malignant oral diseases (leukoplakia, oral submucous fibrosis, lichen planus), oral squamous cell carcinoma (OSCC) and oral submucous fibrosis (OSMF) was shown. Three studies in diabetes patients show a consistent increase of MN frequency reproducing the data available on MN in peripheral lymphocytes which was suggested as a possible tool for screening of metabolic syndromes. Small size studies showed an increase of MN frequency in patients with chronic diseases such as rheumatoid arthritis, Behcet's disease, tonsillitis and renal disease. However higher frequencies of these anomalies were reported in buccal cells from individuals with pathological oral lesions such as dysplastic leukoplakia with respect to the controls suggesting their potential role as complementary markers of cell death and/or malignant transformation. Peculiar cytome profile was observed in Alzheimer patients characterized by a lower frequency of basal cells, condensed chromatin and karyorrhectic cells suggesting alterations in cell kinetics and cell death process indicative of reduced regenerative capacity of epithelial tissue.

CONCLUSION

As has been shown, MN as a biomarker of DNA damage is a very useful tool that can be applied for *in vitro* and *in vivo* models. Along with the cytome the ambiental influence of a single or a mixture of xenobiotics can be measured on the induction of cell death or genotoxic damage. This assay is a very useful and easy way to determine genotoxicans, and cytotoxic agents under laboratory controled conditions as well as the effect of them in the enviroment when it is used in populations or as clinical biomarker to predict genomic inestability and its consequences.

REFERENCES

Abramsson-Zetterberg L, Grawé J, Zetterberg G. 1999. "The micronucleus test in rat erythrocytes from bone marrow, spleen and peripheral blood: the response to low doses of ionizing radiation, cyclophosphamide and vincristine determined by flow cytometry." *Mutation Research/ Fundamental and Molecular Mechanism of Mutagenesis* 423:113-124.

Asano, N, Katsuma Y, Tamura H, Higashikuni N and Hayashi M. 1998. "An automated new technique for scoring the rodent micronucleus assay: computerized image analysis of acridine orange supravitally stained peripheral blood cells." *Mutation Research/Fundamental and Molecular Mechanism of Mutagenesis* 409:149–154.

Boller, K. J., Schmid W. K., 1970. "Chemical mutagenesis in mammals. The Chinese hamster bone marrow as an *in vivo* test system. Hematologic findings after treatment with Trenimon." *Human genetics* 1:35–54.

Bolognesi, C., Knasmueller S., Nersesyan A., Thomas P., Fenech M. 2013. "The HUMNxl scoring criteria for different cell types and nuclear anomalies in the buccal micronucleus cytome assay-An update and expanded photogallery." *Mutation Research/Fundamental Mechanism of Mutagenesis* 753:100-113.

Bolognesi, C., Bonassi S., Knasmueller S., Fenech M., Bruzzone M., Lando C., Ceppi M. 2015. "Clinical application of micronucleus test in exfoliated buccal cells: A systematic review and metanalysis." *Mutation Research/Reviews in Mutation Research* 766:20-31.

Crasta, K., Ganem N., Dagher R., Lantermann A. B., Ivanova, E., Pan Y., Nezi L., Protopopov A., Chowdhury D., Pellman D. 2012. "DNA breaks and chromosome pulverization from errors in mitosis." *Nature* 482:53-60.

Dertinger, S., Torous D., Hayashi M., MacGregor J. 2011. "Flow cytometric scoring of micronucleated erythrocytes: an efficient platform for *assessing in vivo* cytogenetic damage." *Mutagenesis* 26:139-145.

Dertinger, S. D., Ying T., Nowak I., Hyrien O., Sun H., Bemis J. C., Torous D., Keng P., Palis J., and Chen Y. 2007. "Reticulocyte and micronucleated reticulocyte responses to gamma irradiation: dose-response and time-course profiles measured by flow cytometry." *Mutation Research/Genetic Toxicology and Environmental Mutagenesis* 634:119-125.

Evans, H. J., G. J. Neary and F. S. Williamson. 1959. "The Relative Biological Efficiency of Single Doses of Fast Neutrons and Gamma-rays on Vicia Saba Roots and the Effect of Oxygen. Part II. Chromosome Damage: The Production of Micronuclei." *Journal of Radiation Biology and Related Studies in Physics, and Medicine* 1:216-229.

Fenech, M., Kirsch-Volders M., Natarajan, A. T., Surralles J., Crott J. W., Parry J., Norppa H., Eastmond D. A, Tucker J.. D., Thomas P. 2011. "Molecular mechanism of micronucleus, nucleoplasmic bridge and nuclear bud formation in mammalian and human cells." *Mutagenesis* 26:125-132.

Fenech, M. 1993. "The cytokinesis-block micronucleus technique: A detailed description of the method and its application to genotoxicity studies in human populations." *Mutation Research/Fundamental and Molecular Mechanisms of Mutagenesis* 285:35-44.

Fenech, M. 2007. "Cytokinesis-block micronucleus cytome assay." *Nature Protocols 2:*1084-1104.

Gómez-Meda, B. C., G. M. Zúñiga-González, L. V. Sánchez-Orozco, A. L. Zamora-Perez, J. P. Rojas-Ramírez, A. D. Rocha-Muñoz. A. A Guerrero-de León, J. S. Armendáriz-Borunda and M. G Sánchez-Parada. "Buccal micronucleus cytome assay of populations under chronic heavy metal and other metal exposure along the Santiago River, Mexico" *Environmental monitoring Assessment*: 189: 522.

Hayashi, M, Norppa H, Sofuni T and Ishidate M. Jr. 1992. "Flow cytometric micronucleus test with mouse peripheral erythrocytes." *Mutagenesis* 7:257-264.

Hayashi, M., Sofuni T., Ishidate S Jr. 1983. "An application of Acridine Orange fluorescent staining to the micronucleus test." *Mutation Research Letters* 120:241-247.

Hayashi, M. 2016. "The micronucleus test-most widely used *in vivo* genotoxicity test." *Genes and Environment* 38:3818-3823.

Heddle, J. A, Fenech, M., Hayashi M., MacGregor J. 2011. "Reflections on the development of micronucleus assay." *Mutagenesis* 26:3-10.

Holland, N., Bolognesi, C., Kirsch-Volders, M., Bonassi, S., Zeiger, E., Knasmueller, S., Fenech M. 2008. "The micronucleus assay in human buccal cells as a tool for biomonitoring DNA damage: The HUMN project perspective on current status and knowledge gaps." *Mutation Research/Reviews in Mutation Research* 659:93-108.

Hothorn, L. A., Gerhard. D. 2009. "Statistical evaluation of the *in vivo* micronucleus assay." *Archives Toxicology* 83:625-634.

Hovhannisyan, G., Harutyunyan T., Aroutionian. R. 2018. "Micronuclei and What They Can Tell Us in Cytogenetic Diagnostics." *Current Genetic Medicine Reports* 6:144-154.

Hutter, K. J. and Stör M. 1982. "Rapid detection of mutagen induced micronucleated erythrocytes by flow cytometry." *Histochemistry* 75:353-362.

Kiraly, G., Simonyi, A., Turani, M., Juhasz, I., Nagy G., Banfalvi G. 2016. "Micronucleus formation during chromatin condensation and under apoptotic conditions." *Apoptosis* 22:207-219.

Kirsch-Volders, M., Plas, G. Elhajouji A., Lukamowicz, M., Gonzalez L., Vande K., Loock S., Decordier I. 2011. "The *in vitro* MN assay in 2011: origin and fate, biological significance, protocols, high throughput methodologies and toxicological relevance." *Archives of Toxicology* 85:873-899.

Kirsch-Volders, M., Mateuca Raluca A., Roelants M., Tremp A., Zeiger E., Bonassi S., Holland N., Wushou P., Chang, C., Aka V., DeBoeck M, Godderis L., Haufroid V., Ishikawa H., Laffon B., Marcos R., Migliore L., Norppa Hannu, Teixeira J., Zijno A., Fenech M. 2006. "The effects of GSTM1 and GSTT1 polymorphisms on micronucleus frequencies in human lymphocytes *in vivo*." *Cancer Epidemiology Biomarkers and Prevention* 15:1038- 1042.

Kisurina-Evgenieva, O. P., O. I. Sutiagina and G. E. Onishchenko. 2016. "Biogenesis of Micronuclei." *Biochemistry (Moscow)* 81:612-624.

Klumpp, A. Ansel W., Fomin, A., Schnirring S., Pickl C. 2004. "Influence of climatic conditions on the mutations in pollen mother cells of *Tradescantia* clone 4430 an implications for the Trad- MCN bioassay protocol." *Hereditas* 141:142-148.

Krishna, G. and Hayashi M. 2000. "*In vivo* rodent micronucleus assay: protocol, conduct and data interpretation." *Mutation Research/Fundamental and Molecular Mechanisms of Mutagenesis* 455:155-166.

Lee, J., Hung, H. C. Cheng S. J., Chen, Y. J. Chiang C. P., Liu. B. Y. 2006. "Carcinoma and dysplasia in oral leukoplakias in Taiwan: prevalence and risk factors," *Oral Surgery. Oral. Medicine. Oral Pathology. Oral Radiology* 101:472–480.

Li, G., Pan Y., Shuai H., Wei H., Cheng L., Bing-Shuang Z., d Xu-Chen Ma. 2018. "Buccal mucosa cell damage in individuals following dental X-ray examinations." *Scientific Reports* 8:1-7.

Ma T. H., Anderson V. A., Ahmed I. (1982) Environmental Clastogens Detected by Meiotic Pollen Mother Cells of Tradescantia. In: Tice R. R., Costa D. L., Schaich K. M. (eds) Genotoxic Effects of Airborne Agents. *Environmental Science Research*, vol 25. Springer, Boston, MA.

Ma, T. H., Cabrera G. L., Chen R., Gill B. S., Sandhu S. S., Vandenberg A.L. and Salamone M. F. 1994. "Tradescantia micronucleus bioassay." *Mutation Research/Fundamental and Molecular Mechanisms of Mutagenesis* 310:221-230.

Ma, T. H. 1981. "Tradescantia Micronucleus Bioassay and Pollen Tube Chromatid Aberration Test for *in Situ* Monitoring and Mutagen Screening." *Environmental Health Perspectives* 37:85-90.

MacGregor, J. T., Wehr C. M. and Langlois R. G. 1983. "A simple fluorescent staining procedure for micronuclei and RNA in erythrocytes using Hoechst 33258 and pyronin Y." *Mutation Research Letters* 120:269-275.

Misik, M., Krupitza G., Misikova K., Micieta K., Nersesyan A., Kundi M., Knasmueller S. 2016. "The Tradescantia micronucleus assay is a highly sensitive tool for the detection of low levels of radioactivity in environmental samples." *Environmental Pollution* 219:1044-1048.

Morita, T., Shuichi, H., Kenichi M., Akihiro W., Jiro M., Hironao T., Katsuaki Y., Tsuneo H., Masamitsu, H. 2016. "Evaluation of the sensitivity and specificity of *in vivo* erythrocyte micronucleus and transgenic rodent gene mutation test to detect rodent carcinogens." *Mutation Research/Genetic Toxicology and Environmental Mutagenesis* 802:1-29.

Nersesyan, A., Kundi, M., Fenech, M., Bolognesi, C., Misi, M., Wultsch, G., Hartmann M., Knasmueller, S. 2014. "Micronucleus assay with urine derived cells (UDC): A review of its application in human studies investigating genotoxin exposure and bladder cancer risk." *Mutation Research/Reviews in Mutation Research* 762:37-51.

Norppa, H., Ghita, C., Flack, M. 2003. "What do human micronuclei contain?." *Mutagenesis* 18:221-233.

OECD (2014), *Test No. 474: Mammalian Erythrocyte Micronucleus Test*, OECD Publishing, Paris, https://doi.org/10.1787/9789264224292-en.

Patino-Garcia, B., Hoegel, J., Varga, D., Hoehne, M., Michel, I., Jainta, S., Kreienberg, R., Maier C., Vogel, W. 2006. "Scoring variability of micronuclei in binucleate human lymphocytes in a case-control study." *Mutagenesis* 21:191-197.

Randa, El-Zein, Vra, Al. Etzel, C. 2011. "Cytokinesis-blocked micronucleus assay and cancer risk assessment." *Mutagenesis* 26:101-106.

Randa, El-Zein, Michael Fenech, M.,, S, Lòpez, M., Spitz, R., Etzel, C, 2008. "Cytokinesis-Blocked Micronucleus Cytome Assay Biomarkers Identify Lung Cancer Cases Amongst Smokers." *Cancer Epidemiol Biomarkers Preview* 17:1111-1119.

Resende de Morais, C., Boscolli Barbosa Pereira, P., Almeida Sousa, A. Santana V., Vieira Santos Campos, C., Malfinato S., Carvalho, M., Spanó, A., Alves de Rezende A., Bonetti, A. 2019. "Evaluation of the genotoxicity of neurotoxic insecticides using the micronucleus test in Tradescantia pallida." *Chemosphere* 227:371-380.

Romagna, F. and Staniforth C. D. 1989. "The automated bone marrow micronucleus test." *Mutation Research/Fundamental and Molecular Mechanisms of Mutagenesis* 213:91-104.

Rosin, M. Och, R. A. Gatti R. Boder. E. 1989. "Heterogeneity of chromosomal breakage levels in epithelial tissue of ataxia-telangiectasia homozygotes and heterozygotes." *Human Genetics* 83:133-138.

Russo, A, Degrassi, F. 2018. "Molecular cytogenetics of the micronucleus: Still surprising." *Mutation Research/Genetic Toxicology and Environmental Mutagenesis* 836:36-40.

Sara, R, Tiwari R. K., Reddy P. P. and Ahuja Y. R. 2007. "Risk assessment of oral cancer in patients with pre-cancerous states of the oral cavity using micronucleus test and challenge assay." *Oral Oncology* 44:354-360.

Schmid, W. 1975. "The micronucleus test." *Mutation Research/Envirinmental Mutagenesis and Related Subjects* 31:9-15.

Schupp, N., Stopper H., Heidland. A. 2016. "DNA Damage in Chronic Kidney Disease: Evaluation of Clinical Biomarkers." *Oxidative Medicine and Cellular Longevity* 2016:1-10.

Steinkellner, H., Mun-Sik, K., Helma, C., Ecker, S., Ma, T. H., Horak, O., Kundi, M., Knasmuller, S., 1998. "Genotoxic effects of heavy metals: comparative investigation with plant bioassays." *Environmental. Mol. Mutagenics.* 31, 183e191.

Suyama, F., Guimaraes, E. T., Lobo, D. J., Rodrigues, G.S., Domingos, M., Alves E. S., Carvalho H. A., and Saldiva P. H. 2002. "Pollen mother cells of Tradescantia clone 4430 and Tradescantia pallida var. purpurea are equally sensitive to the blastogenic effects of X-rays." *Brazilian Journal of Medical and Biological Research* 35:127-129.

Watson, L. and Dallwitz, M. J. 1994. "The Families of Flowering Plants. Interactive Identification and Information Retrieval on CD-ROM version 1.0 1993, and colour illustrated manual." *Nordic Journal of Botanic* 14:486-486.

Zhang, C., Alexander S., Hauke C., Francis J-, Jackson E., Shinwei L., Meyerson M., Pellman D. 2015. "Chromothripsis from DNA damage in micronuclei." *Nature* 522:179-205.

BIOGRAPHICAL SKETCH

Alejandra Hernández-Ceruelos

Affiliation: Cuerpo Académico de Salud Pública, Medical School, Instituto de Ciencias de la Salud Universidad Autónoma del Estado de Hidalgo, México

Education: PhD Chemical biological Science

Business Address: 5a Etapa, Área Académica de Medicina, Instituto de Ciencias de la Salud, Universidad Autónoma del Estado de Hidalgo,

Carretera Pachuca-Actopan camino a Tilcuautla s/n Pueblo San Juan Tilcuautla, 42160 Hidalgo, México.

Research and Professional Experience: Genetic toxicology, human cytogenetics, antimutagenesis and chemoprotection of natural products, biomarkers, public health.

Professional Appointments: Full time Professor Researcher

Honors: PhD with honorific mention, Master's degree with honorific mention, CONACYT scholar grant

Publications from the Last 3 Years:

Hernandez- Ceruelos A., Romero- Quezada L., Rulvalcaba Ledezma J., López Conteras L. 2019. Therapeutic uses of metronidazole an tis side effects: an update. *European Review for Medican and Pharmacological Sciences.* 23: 401-408. doi: 10.26355/eurrev_201901_16788.

Ruvalcaba-Ledezma J., Sanchez-Gervacio B., Hernandez-Cruz A., Lopez Contreras L., Hernandez-Ceruelos A., Reynoso-Vazquez J. 2018. Asociación entre medio ambiente y salud pública: El caso del incendio del Relleno Sanitario en Mineral de la Reforma, Hidalgo, México. *Salud y Educación: boletín científico de ciencias de la salud ICSA*, 13: 96-98 doi: https://doi.org/10.29057/icsa.v7i13.

Vazquez Alvarado P, Herandez Ceruelos A., Sergio Muñoz Juarez S. 2016. Fluoruros, Medio Ambiente y Salud. *Revista de Educación Cooperación y Bienestar Social* 9: 43-52.

Sergio Muñoz-Juárez

Affiliation: Research department, Hospital General de Pachuca, Secretaría de Salud de Hidalgo, México

Education: PhD in Public Health Science

Business Address: Research department, Hospital General de Pachuca, Carretera Pachuca-Tulancingo 101, Nueva Francisco I Madero, La Hacienda, 42070 Pachuca de Soto, Hidalgo, México

Research and Professional Experience: Public health, Biostatistic expert

Professional Appointments: Head of the research department, CONACYT scholar grant

Honors: Chairman of the Research Ethics Committee at Pachuca's General Hospital

Publications from the Last 3 Years:

Vazquez Alvarado P, Herandez Ceruelos A., Sergio Muñoz Juarez S. 2016. Fluoruros, Medio Ambiente y Salud. *Revista de Educación Cooperación y Bienestar Social* 9: 43-52.

Fernando Vázquez-Rivera

Affiliation: Medical School, Instituto de Ciencias de la Salud Universidad Autónoma del Estado de Hidalgo, México

Education: Medical doctor

Business Address: 5a Etapa, Área Académica de Medicina, Instituto de Ciencias de la Salud, Universidad Autónoma del Estado de Hidalgo, Carretera Pachuca-Actopan camino a Tilcuautla s/n Pueblo San Juan Tilcuautla, 42160 Hidalgo, México.

Professional Appointments: assistant of professor

Jesús Ruvalcaba-Ledezma

Affiliation: Cuerpo Académico de Salud Pública, Medical School, Instituto de Ciencias de la Salud Universidad Autónoma del Estado de Hidalgo, México

Education: PhD in Public Health Science

Business Address: 5a Etapa, Área Académica de Medicina, Instituto de Ciencias de la Salud, Universidad Autónoma del Estado de Hidalgo, Carretera Pachuca-Actopan camino a Tilcuautla s/n Pueblo San Juan Tilcuautla, 42160 Hidalgo, México.

Research and Professional Experience: Public Health, biomarkers

Professional Appointments: Full time Professor Researcher

Honors: CONACYT scholar grant, AMMFEN Award for Researcher of the year 2008

Publications from the Last 3 Years:

Hernandez- Ceruelos A., Romero- Quezada L., Rulvalcaba Ledezma J., López Conteras L. 2019. Therapeutic uses of metronidazole an tis side effects: an update. *European Review for Medican and Pharmacological Sciences.* 23: 401-408. DOI: 10.26355/eurrev_ 201901_16788.

Ruvalcaba-Ledezma J., Sanchez-Gervacio B., Hernandez-Cruz A., Lopez Contreras L., Hernandez-Ceruelos A., Reynoso-Vazquez J. 2018. Asociación entre medio ambiente y salud pública: El caso del incendio del Relleno Sanitario en Mineral de la Reforma, Hidalgo, México. *Salud y Educación: boletín científico de ciencias de la salud ICSA*, 13: 96-98 doi: https://doi.org/10.29057/icsa.v7i13.

Luilli López-Contreras

Affiliation: Cuerpo Académico de Salud Pública, Medical School, Instituto de Ciencias de la Salud Universidad Autónoma del Estado de Hidalgo, México

Education: PhD Infectomic and Molecular Pathogenesis

Business Address: 5a Etapa, Área Académica de Medicina, Instituto de Ciencias de la Salud, Universidad Autónoma del Estado de Hidalgo, Carretera Pachuca-Actopan camino a Tilcuautla s/n Pueblo San Juan Tilcuautla, 42160 Hidalgo, México.

Research and Professional Experience: Biomarkers

Professional Appointments: Full time Professor Researcher

Honors: CONACYT scholar grant, Master´s degree with honorific mention

Publications from the Last 3 Years:

Hernandez- Ceruelos A., Romero- Quezada L., Rulvalcaba Ledezma J., López Conteras L. 2019. Therapeutic uses of metronidazole an tis side effects: an update. *European Review for Medican and Pharmacological Sciences*. 23: 401-408. DOI: 10.26355/eurrev_ 201901_16788.

Ruvalcaba-Ledezma J., Sanchez-Gervacio B., Hernandez-Cruz A., Lopez Contreras L., Hernandez-Ceruelos A., Reynoso-Vazquez J. 2018. Asociación entre medio ambiente y salud pública: El caso del incendio del Relleno Sanitario en Mineral de la Reforma, Hidalgo, México. *Salud y Educación: boletín científico de ciencias de la salud ICSA*, 13: 96-98 doi: https://doi.org/10.29057/icsa.v7i13.

In: Micronucleus Assay: An Overview
Editor: Robert C. Cole

ISBN: 978-1-53616-678-1
© 2020 Nova Science Publishers, Inc.

Chapter 3

PIVOTAL ROLE OF MICRONUCLEUS TEST IN DRUG DISCOVERY

Hasan Türkez[*], *Mehmet Enes Arslan and Adil Mardinoğlu*

[1]Department of Pharmacy, University "G. d'Annunzio" of Chieti-Pescara, Chieti Scalo (CH), Italy
Department of Molecular Biology and Genetics, Faculty of Science, Erzurum Technical University, Erzurum, Turkey
[2]Department of Molecular Biology and Genetics, Faculty of Science, Erzurum Technical University, Erzurum, Turkey
[3]Science for Life Laboratory, KTH - Royal Institute of Technology, Stockholm, Sweden, Faculty of Dentistry, Oral and Craniofacial Sciences, Centre for Host-Microbiome Interactions, King's College London, London, UK
Department of Chemical and Biological Engineering, Chalmers University of Technology, Gothenburg, Sweden

---

[*] Corresponding Author's Email: hasanturkez@erzurum.edu.tr.

Abstract

Early detection of adverse effects of novel compunds during drug discovery and development most probably reduce late stage failures, expenses and exertions for candidate drugs. Although the micronucleus (MN) test is one of the oldest techniques used in biochemical sciences for drug discovery. Flexibility of the technique for both *in vitro* and *in vivo* applications and practicability for large scale samples in short time make the MN test an inevitable tool for chemical trails. Drug studies require a formulation that provides the highest exposure to detect clastogenic and aneugenic activities and thus analysis makes it possible to get the necessary safety margin to support clinical trials. The MN test is one of the most important tools of the genotoxicity test battery in preclinical studies to identify negative effects of compounds that induce numerical and structural chromosome alterations in wide spectrum concentrations. The MN assay can be applied various cell types in different protocols. For instance; the most recommended protocols are bone the marrow micronucleus analysis and the *in vivo* mammalian erythrocyte precursor assay. Also, the rodent ovary cells validation test is a very powerful approach to analyse side effects of a compound. Beside cell types, detection systems can be constituted to obtain a high throughput screening such as integrating flow cytometry analysis into the MN inspections. Since a new compound is needed for such an assay, the MN test can assess abnormalities earlier in the drug discovery pipeline, making structure/genotoxicity connection a possible system for drug characterization.

Keywords: micronucleus, drug discovery, genotoxicity testing, drug safety

Introduction

Before a new drug is released into the market, it must be investigated throughout its chemico-biological properties, efficacy and safety (Purves et al. 1995). When a new drug initially discovered, financial costs start to increase exponentially through improvement phase to market registration. Moreover, drug discovery process is constantly shifting because of changes in technology innovations which makes it harder to find a reliable path in analysis of molecule as a candidate drug. A great number of molecules can be synthesized rapidly by development of chemistry techniques and the diversity of these compounds need to be investigated comprehensively by biological assays in order to constitute safety reports (Kennedy 1997). Genotoxicity analysis is a tool for investigating hazardous effects of a new chemical entities in the aspect of DNA damage such as structural chromosomal aberration, gene mutation, numerical and recombination changes. Also, these alterations can affect future generations through heritable changes on germ cells (Madle et al. 1987; Wassom 1989). On the other hand, somatic mutations have been shown to result in carcinogenesis and effects millions of life severely. These tests are used for investigating genotoxicity and carcinogenicity of a compound to assess possibility of a new drug discovery (Tennant et al. 1987).

Successful drug discovery researches generally depend on early assessment of *in vivo/in vitro* toxicological properties of candidate drugs related with a chosen therapeutic target. In some aspects this aim can be accomplished with simple strategy formulations while others need high throughput systems and technological tools to inspect candidate drug properties (Kwong et al. 2011). Last decade, a great part of the literature has been deliberating about strategies for discovery and development of interactions with the pharmaceutical industry. As respectively, drug development scientists and drug discovery company's collaborations have been increased significantly. Despite these progresses additions of new drugs to the market hasn't been increased and discovery costs have been rising continuously. These consequences are resulted from inability to understand preclinical toxicological properties of a molecule with high proofreading rate. Minimizing time and cost consuming toxicological approaches that can be used in preclinical application in every phase of *in vitro* and *in vivo* drug discovery analysis would overcome these obstacles. To achieve this objective it is inevitable to switch into standardized techniques which are easily used in toxicology researches and also give high productivity results in each analysis. The micronucleus (MN) test can be one of the high throughput screening and analysis technique used in both *in vitro* and *in vivo* analysis to cope with many obstacles related with a new drug discovery (Hayashi 2016; Miller 1998).

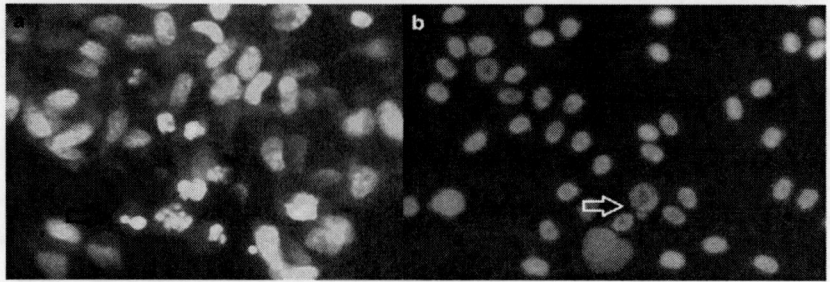

Figure 1. Possible results observed from *in vitro* (a) and *in vivo* (b) MN test by using hoechst 33342 fluorescent staining. **a**- SHSY-5Y neuroblastoma cell line culture, **b**- Rainbow Trout (Oncorhynchus mykiss) fish blood smear slide.

At first the micronucleus was investigated in cats and rats blood by Howell and Jolly in the end of the 19th century. Analysis showed that small inclusions which were called "Howell-Jolly body" could be formed in severe anaemia patients and these terms were the first micronucleus descriptions in the literature (Hayashi 2016). When a genotoxic agent exposed to an organism, toxic potential of the agent can be assed via analysing extracellular bodies which are damaged chromosomal parts or micronucleus sites (Figure 1). One of the essential parts in a genotoxicity test is chromosomal level DNA damage studies because mutations in chromosomes has a crucial role in many cellular events as carcinogenesis. The MN assay is become a very important procedure in toxicology studies because this assay is generally used as an alternative analysis system to the chromosomal aberration test. MN scoring has been performed by using different types of staining techniques such as giemsa colorimetric and hoechst 33342 fluorescent staining assays (Turkez et al. 2017; Jain et al. 2018).

IN VIVO RODENT MICRONUCLEUS ASSAY

One of the most widely used methods for detecting genotoxicity has been *in vivo* rodent MN assay. The primary *in vivo* test in genotoxicity and drug discovery analysis is MN induction evaluation and this method is approved by health institutes around the world to comprise in safety assessments of products. Aneugenicity and clastogenicity can be detected significantly and comprehensively when the assay performed appropriately (Krishna and Hayashi 2000). When rodent cells analysed against a genotoxic agent, it was found to increase micronucleated erythrocytes and cause aberrations in chromosomal structures (Heddle et al. 1983; Heddle and Carrano 1977). The MN tests mostly focus on micronucleated immature erythrocytes (MIE) to increase sensitivity and facilitate scoring in drug discovery researches. Recently, the rodent erythrocyte MN test has been replaced the bone marrow chromosome aberration analysis because it has become less technically challenging, less labour-intensive and it makes possible to detect all chromosomal breakages due to toxic agents (Proudlock et al. 1997). Moreover, it is very critical to use other tissues than bone marrow to investigate risk factors by using the MN assay for shedding light into modes of actions and systemic genotoxic properties of newly found drugs or chemicals. Although, it is important to analyse many different tissue types for toxicological studies, majority of tissues need to validated and standardised protocols and techniques for toxicological applications to get reliable results (Morita et al. 2011). Until now many different tissue types have been used to detect MN formations in safety assessments for drug discovery and toxicological studies (Figure 2). One of the most important tissues other than bone marrow is the liver for studying toxicological properties of a molecule. However, liver cells replicating capacity is highly reduced in adult animals, cell growth of this tissues needs to be stimulated by application of hepatotoxicants or partial hepatectomy (Braithwaite and Ashby 1988; Tates et al. 1983; Tates et al. 1980; Uryvaeva and Delone 1995). This type of problems can be solved by using after birth rodents up to 5-6 weeks old. Usage of young mice or rats can get rid of chemical or physical treatments for cell division stimulation for the MN frequency analysis (Roy et al. 2005; Suzuki et al. 2005; Udroiu et al. 2006). Also, colon and intestinal epithelium are very important target tissues in rodent models for drug safety studies because micronucleus tests in intestine and colon tissues applicable for risk assessment and hazard level identification of a molecule that enter the human body through the food system (Çelik et al.

2005; Poul et al. 2009). In addition to these examples many other tissues have been identified to use for the micronucleus frequency evaluation in safety assessments during drug discoveries such as skin, stomach, bladder, spleen, lung and mucosa tissues.

FLOW CYTOMETRIC MICRONUCLEUS ASSAY

Further development of micronucleus scoring techniques would increase utility of the assay in a more high efficient way and this system can be used in the early drug discovery for faster and optimized analyses. Different types of flow cytometric micronucleus methods are developed for scoring micronuclei in various cell culture models but, a disadvantage of this system is difficulties in distinguishing MN bodies from necrotic and apoptotic chromatins (Avlasevich et al. 2006). Furthermore, the most advanced procedure was constituted for a high throughput MN scoring in the flow cytometric (FCM) analysis by Nüsse and friends (Nüsse et al. 1994; Nüsse and Kramer 1984). FCM-gating and staining genotoxicity procedures were proposed on a mouse lymphoblastoid cell line and according to the results various toxic molecules gave similar results in both FCM measurements and the standard microscopic scoring (Moore et al. 2003). A study was claimed that the *in vivo* micronucleus test with flow cytometry on rat peripheral blood samples gave sensitive results for analysis of different toxic compounds and also it was proposed that only small amount of blood samples can be effectively used for toxicity studies routinely (Cammerer et al. 2007). Moreover, it was suggested that the flow cytometric analysis can significantly increase the information composition gathered from *in vitro* micronucleus screening assays. The interpreted data from different flow cytometric studies were confirmed to be used for identifying and eliminating irrelevant positive results (Bryce et al. 2007).

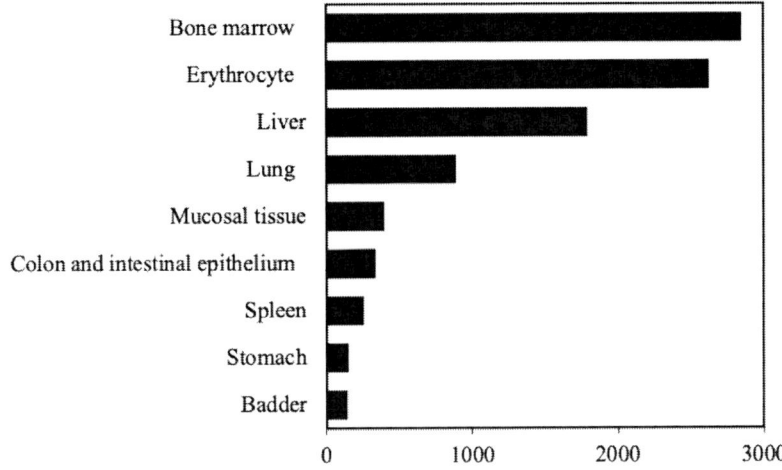

Figure 2. Different tissue types used in micronucleus analysis was shown according to usage frequency in the literature. (Data gathered from: https://www.ncbi.nlm.nih.gov/pubmed).

AUTOMATED ANALYSIS IN THE MICRONUCLEUS ASSAY

The micronucleus analysis is one of the best choices for early drug discovery assessment in constructing safety structure reports. But, the laborious nature of the MN assay where at least 2000 cells need to be investigated for reaching significant statistical information makes the technique difficult to use in some cases. In addition to flow cytometric approaches, different automated systems that replace the boring and time-consuming visual cellular analysis procedures have been debated (Frieauff et al. 2013). One of the first automated micronucleus analysing technology was designed on erythrocytic cells as a computerized image investigating system. Also, this technique was applied successfully on peripheral blood and bone marrow specimens from various species. The fundamental step for this approach is preparing "flat" cell formations on polylysine-coated slides by cytocentrifugation for improving slide quality (Romagna and Staniforth 1989). Different staining, fixation and preparation techniques were determined in the first discovered automated detection system for analysing binucleated cells and MNs using a microscopic image analysis. This system was reported to detect correctly both binucleated cells and micronucleus in range of 50% to 80% which wasn't actually close to microscopic investigations (Tates et al. 1990; Castelain et al. 1993; Vral et al. 1994; Böcker et al. 1995; Decordier et al. 2011b). Recent studies put forth that the Metasystems Metafer MNscore system produced a higher correlation between the automatic and the visual counting and according to the system algorithm if a two similar nuclei are located at a defined region, it can be count as a binuclear cell. In addition, the system was claimed to be used to analyse nonflourescent/colorimetric stained samples for long term analysis. Researches showed that different modifications on slide preparation increased micronucleus detection ratio up to 87% in different samples by using the same Metafer system (Schunck et al. 2004; Varga 2004; Willems et al. 2010). A final form of the automated system RABiT-II was proposed to give nearly perfect results in analysing micronucleus detection and counting. In this study, the automated cytokinesis block micronucleus (CBMN) assay was demonstrated to be automated in a high throughput format by using different robotic systems using triage radiation biodosimetry in standardized multiwell plates (Repin et al. 2019).

MICRONUCLEUS ASSAY AND LABELING OF CENTROMERES WITH FISH TECHNIQUE

The CBMN assay has been used in drug discovery processes for biomonitoring toxicological properties of clastogenic and aneugenic events on human tissues for many years. This technique has become a standard *in vivo*/*ex vivo* procedure for comprehensive assessment of cytostasis, cytotoxicity and genetic damage on human lymphocytes. Also, drug applications and cancer occurrence risk relationships have been demonstrated in the aspect of the frequency of MNs by analysing cytokinesis-blocked lymphocytes. The CBMN assay has a distinguishing advantage on investigating and differentiating both DNA breakages and chromosome losses by consolidation of the assay in the Fluorescence *in situ* hybridization (FISH) technique with the centromeric labelling (Decordier et al. 2011a). Micronucelus structures resulted from underdeveloped chromosomes can be determined by the kinetochore protein labelling on centromeres or addition of centromeric DNA tags for FISH analysis. Kinetochore proteins or centromeric DNA regions don't exist on the MN formations and these features can be characterized to confirm toxicological issues (Albertini et al. 2000). Some studies have shown that the MN analysis, together with FISH, can be used *in vitro* to display thresholds that specifically bind to b-tubulin and inhibit tubulin polymerization, such ascarbendazim or nocodazole (Elhajouji et al. 1997). For further development of the MN analysis combination with FISH can be performed by integrating automated scoring systems into the process. Some researches that previously touched on in this chapter have been improving the automated MN scoring validation and recent progress would permit researchers to get faster, high throughput and more reliable safety structure analysis of new drug discoveries (Wilde et al. 2019).

CONCLUSION

Due to its accuracy, simplicity and multi-potentiality in preclinical drug discovery studies micronucleus analysis has become inevitable tool for the predictive toxicological analysis system for both *in vivo* and *in vitro* researches. Application of the assay in a broad spectrum of research areas such as genotoxicity investigations in chemical exposures, pharmacogenomics, toxicogenomics, nutrigenomics, ecotoxicology and medicine expand use of the technique in safety report constitution investigation scope. The most common application of the MN assay has consisted of the *in vivo* rodent toxicological analysis. Different tissue types of rodent models can be used to investigate various toxicological properties and plasticity of the assay increases the applications in early drug research areas from erythrocytes to lung studies (Gebel et al. 1997; Hayashi et al. 2000; Kirkland et al. 2011; Puteri et al. 2015). Integration of innovative technologies into the MN assay makes the procedure faster and more reliable for early drug safety studies which show unlimited potential of the assay. Recent studies combining flow cytomety and micronucleus analysis have been producing high throughput and comprehensive information about molecules in toxicological studies which could possibly increase number of novel drugs introductions (Rodrigues et al. 2016; Di Bucchianico et al. 2017; Lebedová et al. 2018; Wang et al. 2019). Robotic and automated scoring software systems have been established for the MN assay in different tissue types. By incorporating computational processes into toxicological investigations can eliminate traditional MN assay's time consuming and exhausting disadvantages (Roemer et al. 2015; Depuydt et al. 2017; Repin et al. 2017). Moreover, various commonly used biological procedures can enhance use of the MN assay in different conditions of drug discovery studies. For instance, FISH technique can strengthen preclinical drug analysis on the molecular impact from DNA breakages to chromosomal loss mutations and this technical integration can widen perspective of newly established researches (Yadav 2015; Hovhannisyan et al. 2017; Petibone and Ding 2018).

REFERENCES

Albertini, R. J., Anderson, D., Douglas, G. R., Hagmar, L., Hemminki, K. & Merlo, F. (2000). IPCS guidelines for the monitoring of genotoxic effects of carcinogens in humans. *Mutation Reserach*, *463*, 111-172.

Avlasevich, S. L., Bryce, S. M., Cairns, S. E. & Dertinger, S. D. (2006). *In vitro* micronucleus scoring by flow cytometry: Differential staining of micronuclei versus apoptotic and necrotic chromatin enhances assay reliability. *Environ. Mol. Mutagen.*, *47*, 56-66.

Böcker, W., Müller, W. U. & Streffer, C. (1995). Image processing algorithms for the automated micronucleus assay in binucleated human lymphocytes. *Cytometry*, *19*, 283-294.

Braithwaite, I. & Ashby, J. (1988). A non-invasive micronucleus assay in the rat liver. *Mutat. Res. Mutagen. Relat. Subj.*, *203*, 23-32.

Bryce, S. M., Bemis, J. C., Avlasevich, S. L. & Dertinger, S. D. (2007). *In vitro* micronucleus assay scored by flow cytometry provides a comprehensive evaluation of cytogenetic damage and cytotoxicity. *Mutat. Res. Toxicol. Environ. Mutagen.*, *630*, 78-91.

Cammerer, Z., Elhajouji, A. & Suter, W. (2007). *In vivo* micronucleus test with flow cytometry after acute and chronic exposures of rats to chemicals. *Mutat. Res. Toxicol. Environ. Mutagen.*, *626*, 26-33.

Castelain, P., Van Hummelen, P., Deleener, A. & Kirsch-Volders, M. (1993). Automated detection of cytochalasin-B blocked binucleated lymphocytes for scoring micronuclei. *Mutagenesis*, *8*, 285-293.

Çelik, A., Mazmanci, B., Çamlica, Y., Aşkin, A. & Çömelekoğlu, Ü. (2005). Induction of micronuclei by lambda-cyhalothrin in Wistar rat bone marrow and gut epithelial cells. *Mutagenesis*, *20*, 125-129.

Decordier, I., Mateuca, R. & Kirsch-Volders, M. (2011a). Micronucleus assay and labeling of centromeres with FISH technique. *Methods Mol Biol.*, *691*, 115-136.

Decordier, I., Papine, A., Vande Loock, K., Plas, G., Soussaline, F. & Kirsch-Volders, M. (2011b). Automated image analysis of micronuclei by IMSTAR for biomonitoring. *Mutagenesis*, *26*, 163-168.

Depuydt, J., Baeyens, A., Barnard, S., Beinke, C., Benedek, A. & Beukes, P. (2017). RENEB intercomparison exercises analyzing micronuclei (Cytokinesis-block Micronucleus Assay). *Int. J. Radiat. Biol.*, *93*, 36-47.

Di Bucchianico, S., Cappellini, F., Le Bihanic, F., Zhang, Y., Dreij, K. & Karlsson, H. L. (2017). Genotoxicity of TiO 2 nanoparticles assessed by mini-gel comet assay and micronucleus scoring with flow cytometry. *Mutagenesis*, *32*, 127-137.

Elhajouji, A., Tibaldi, F. & Kirsch-Volders, M. (1997). Indication for thresholds of chromosome non-disjunction versus chromosome lagging induced by spindle inhibitors *in vitro* in human lymphocytes. *Mutagenesis*, *12*, 133-140.

Frieauff, W., Martus, H. J., Suter, W. & Elhajouji, A. (2013). Automatic analysis of the micronucleus test in primary human lymphocytes using image analysis. *Mutagenesis*, *28*, 15-23.

Gebel, T., Kevekordes, S., Pav, K., Edenharder, R. & Dunkelberg, H. (1997). *In vivo* genotoxicity of selected herbicides in the mouse bone-marrow micronucleus test. *Arch. Toxicol.*, *71*, 193-197.

Hayashi, M. (2016). The micronucleus test-most widely used *in vivo* genotoxicity test. *Genes Environ.*, *38*, 18.

Hayashi, M., MacGregor, J. T., Gatehouse, D. G., Adler, I. D., Blakey, D. H. & Dertinger, S. D. (2000). *In vivo* rodent erythrocyte micronucleus assay. II. Some aspects of protocol design including repeated treatments, integration with toxicity testing, and automated scoring. *Environ. Mol. Mutagen.*, *35*, 234-252.

Heddle, J. A. & Carrano, A. V. (1977). The DNA content of micronuclei induced in mouse bone marrow by γ-irradiation: evidence that micronuclei arise from acentric chromosomal fragments. *Mutat. Res. Mol. Mech. Mutagen.*, *44*, 63-69.

Heddle, J. A., Hite, M., Kirkhart, B., Mavournin, K., MacGregor, J. T. & Newell, G. W. (1983). The induction of micronuclei as a measure of genotoxicity. *Mutat. Res. Genet. Toxicol.*, *123*, 61-118.

Hovhannisyan, G., Harutyunyan, T. & Liehr, T. (2017). Micronucleus FISH. Thomas Liehr (ed.), Fluorescence *In Situ* Hybridization (FISH), Springer Protocols Handbooks, Springer, Berlin, Heidelberg, 379-383.

Jain, A. K., Singh, D., Dubey, K., Maurya, R., Mittal, S. & Pandey, A. K. (2018). Models and methods for *in vitro* toxicity. *In Vitro Toxicol.*, 45-65.

Kennedy, T. (1997). Managing the drug discovery/development interface. *Drug Discov. Today.*, 436-444.

Kirkland, D., Reeve, L., Gatehouse, D. & Vanparys, P. (2011). A core *in vitro* genotoxicity battery comprising the Ames test plus the *in vitro* micronucleus test is sufficient to detect rodent carcinogens and *in vivo* genotoxins. *Mutat. Res. Toxicol. Environ. Mutagen.*, *721*, 27-73.

Krishna, G. & Hayashi, M. (2000). *In vivo* rodent micronucleus assay: protocol, conduct and data interpretation. *Mutat. Res. Mol. Mech. Mutagen.*, *455*, 155-166.

Kwong, E., Higgins, J. & Templeton, A. C. (2011). Strategies for bringing drug delivery tools into discovery. *Int. J. Pharm.*, *412*, 1-7.

Lebedová, J., Hedberg, Y. S., Odnevall, W. I. & Karlsson, H. L. (2018). Size-dependent genotoxicity of silver, gold and platinum nanoparticles studied using the mini-gel comet assay and micronucleus scoring with flow cytometry. *Mutagenesis, 33,* 77-85.

Madle, S., Korte, A. & Bass, R. (1987). Experience with mutagenicity testing of new drugs: viewpoint of a regulatory agency. *Mutat. Res. Mutagen. Relat. Subj., 182,* 187-192.

Miller, B. (1998). Evaluation of the *in vitro* micronucleus test as an alternative to the *in vitro* chromosomal aberration assay: position of the GUM working group on the *in vitro* micronucleus test. *Mutat. Res. Mutat. Res., 410,* 81-116.

Moore, M. M., Honma, M., Clements, J., Bolcsfoldi, G., Cifone, M. & Delongchamp, R. (2003). Mouse lymphoma thymidine kinase gene mutation assay: International workshop on genotoxicity tests workgroup report-Plymouth, UK 2002. *Mutat. Res. Toxicol. Environ. Mutagen., 540,* 127-140.

Morita, T., MacGregor, J. T. & Hayashi, M. (2011). Micronucleus assays in rodent tissues other than bone marrow. *Mutagenesis, 26,* 223-230.

Nüsse, M., Beisker, W., Kramer, J., Miller, B. M., Schreiber, G. A. & Viaggi, S. (1994). Measurement of micronuclei by flow cytometry. *Methods Cell Biol.,* 149-158.

Nüsse, M. & Kramer, J. (1984). Flow cytometric analysis of micronuclei found in cells after irradiation. *Cytometry, 5,* 20-25.

Petibone, D. M. & Ding, W. (2018). Fluorescence *in situ* hybridization in genotoxicity testing. *Mutagen. Assays Appl.,* 265-286.

Poul, M., Jarry, G., Elhkim, M. O. & Poul, J. M. (2009). Lack of genotoxic effect of food dyes amaranth, sunset yellow and tartrazine and their metabolites in the gut micronucleus assay in mice. *Food Chem. Toxicol., 47,* 443-448.

Proudlock, R. J., Statham, J. & Howard, W. (1997). Evaluation of the rat bone marrow and peripheral blood micronucleus test using monocrotaline. *Mutat. Res. Toxicol. Environ. Mutagen., 392,* 243-249.

Purves, D., Harvey, C., Tweats, D. & Lumley, C. E. (1995). Genotoxicity testing: current practices and strategies used by the pharmaceutical industry. *Mutagenesis, 10*, 297-312.

Puteri, M. B., Noor, J. & Muhammad, H. (2006). The *in vivo* rodent micronucleus assay of Kacip Fatimah (Labisia pumila) extract. *Trop Biomed., 23*, 214-219.

Repin, M., Pampou, S., Garty, G. & Brenner, D. J. (2019). RABiT-II: A fully-automated micronucleus assay system with shortened time to result. *Radiation Research, 191*, 232-238.

Repin, M., Pampou, S., Karan, C., Brenner, D. J. & Garty, G. (2017). RABiT-II: Implementation of a high-throughput micronucleus biodosimetry assay on commercial biotech robotic systems. *Radiation Research, 187*, 502-508.

Rodrigues, M. A., Probst, C. E., Beaton-Green, L. A. & Wilkins, R. C. (2016). Optimized automated data analysis for the cytokinesis-block micronucleus assay using imaging flow cytometry for high throughput radiation biodosimetry. *Cytom. Part A., 89*, 653-662.

Roemer, E., Zenzen, V., Conroy, L. L., Luedemann, K., Dempsey, R. & Schunck, C. (2015). Automation of the *in vitro* micronucleus and chromosome aberration assay for the assessment of the genotoxicity of the particulate and gas–vapor phase of cigarette smoke. *Toxicol. Mech. Methods, 25*, 320-333.

Romagna, F. & Staniforth, C. (1989). The automated bone marrow micronucleus test. *Mutat. Res. Mol. Mech. Mutagen., 213*, 91-104.

Roy, S. K., Thilagar, A. K. & Eastmond, D. A. (2005). Chromosome breakage is primarily responsible for the micronuclei induced by 1,4-dioxane in the bone marrow and liver of young CD-1 mice. *Mutat. Res. Toxicol. Environ. Mutagen, 586*, 28-37.

Schunck, C., Johannes, T., Varga, D., Lörch, T. & Plesch, A. (2004). New developments in automated cytogenetic imaging: unattended scoring of dicentric chromosomes, micronuclei, single cell gel electrophoresis, and fluorescence signals. *Cytogenet. Genome Res., 104*, 383-389.

Suzuki, H., Ikeda, N., Kobayashi, K., Terashima, Y., Shimada, Y. & Suzuki, T. (2005). Evaluation of liver and peripheral blood micronucleus assays with 9 chemicals using young rats. *Mutat. Res. Toxicol. Environ. Mutagen*, *583*, 133-145.

Tates, A. D., Neuteboom, I., Hofker, M. & den Engelse, L. (1980). A micronucleus technique for detecting clastogenic effects of mutagens/carcinogens (DEN, DMN) in hepatocytes of rat liver *in vivo*. *Mutat. Res. Mutagen. Relat. Subj.*, *74*, 11-20.

Tates, A. D., Neuteboom, I., de Vogel, N. & den Engelse, L. (1983). The induction of chromosomal damage in rat hepatocytes and lymphocytes I. Time-dependent changes of the clastogenic effects of diethylnitrosamine, dimethylnitrosamine and ethyl methanesulfonate. *Mutat. Res. Mol. Mech. Mutagen.*, *107*, 131-151.

Tates, A. D., Van Welie, M. T. & Ploem, J. S. (1990). The present state of the automated micronucleus test for lymphocytes. *Int. J. Radiat. Biol.*, *58*, 813-825.

Tennant, R., Margolin, B., Shelby, M., Zeiger, E., Haseman, J. & Spalding, J. (1987). Prediction of chemical carcinogenicity in rodents from *in vitro* genetic toxicity assays. *Science*, *236*, 933-941.

Turkez, H., Arslan, M. E. & Ozdemir, O. (2017). Genotoxicity testing: progress and prospects for the next decade. *Expert Opin. Drug Metab. Toxicol.*, *13*, 1-10.

Udroiu, I., Ieradi, L. A., Cristaldi, M. & Tanzarella, C. (2006). Detection of clastogenic and aneugenic damage in newborn rats. *Environ. Mol. Mutagen.*, *47*, 320-324.

Uryvaeva, I. V. & Delone, G. V. (1995). An improved method of mouse liver micronucleus analysis: an application to age-related genetic alteration and polyploidy study. *Mutat. Res. Mutagen. Relat. Subj.*, *334*, 71-80.

Varga, D. (2004). An automated scoring procedure for the micronucleus test by image analysis. *Mutagenesis*, *19*, 391-397.

Vral, A., Verhaegen, F., Thierens, H. & Ridder, L. D. (1994). The *in vitro* cytokinesis-block micronucleus assay: a detailed description of an improved slide preparation technique for the automated detection of micronuclei in human lymphocytes. *Mutagenesis*, *9*, 439-443.

Wang, Q., Rodrigues, M. A., Repin, M., Pampou, S., Beaton-Green, L. A. & Perrier, J. (2019). Automated triage radiation biodosimetry: integrating imaging flow cytometry with high-throughput robotics to perform the cytokinesis-block micronucleus assay. *Radiation Research*, *191*, 342.

Wassom, J. S. (1989). Origins of genetic toxicology and the environmental mutagen society. *Environ. Mol. Mutagen.*, *14*, 1-6.

Wilde, S., Queisser, N., Holz, C., Raschke, M. & Sutter, A. (2019). Differentiation of aneugens and clastogens in the *in vitro* micronucleus test by kinetochore scoring using automated image analysis. *Environ. Mol. Mutagen.*, *60*, 227-242.

Willems, P., August, L., Slabbert, J., Romm, H., Oestreicher, U. & Thierens, H. (2010). Automated micronucleus (MN) scoring for population triage in case of large-scale radiation events. *Int. J. Radiat. Biol.*, *86*, 2-11.

Yadav, A. S. (2015). Buccal micronucleus cytome assay- A biomarker of genotoxicity. *J. Mol. Biomark. Diagn.*, *6*, 236.

BIOGRAPHICAL SKETCH

Name: Prof. Dr. Hasan Türkez

Affiliation: Department of Molecular Biology and Genetics, Faculty of Science, Erzurum Technical University, Erzurum, Turkey

Education:

Graduation Date	Degree	University/Faculty/Department
2013-2017	Second Doctorate	Department of Pharmacy, "G. D'annunzio" University of Chieti-Pescara
2004-2007	Doctorate	Ataturk University, Faculty of Science, Department of Biology
2001-2004	Master	Ataturk University, Faculty of Science, Department of Biology
1997-2001	Bachelor's degree	Ataturk University, Faculty of Education, Department of Biology

Business Address: Erzurum Technical University, 25050, Yakutiye, Erzurum, Turkey.

Research and Professional Experience:

- Drug Research and Development,
- Mutagenicity and Carcinogenicity Testing,
- Neurodegeneration and Related Mechanisms
- Nanomedicine-Nanobiotechnology,
- Cancer Biology, Neurotoxicity,
- Mammalian Cell and Organ Culture Techniques

Professional Appointments:
WHO: Toxicology and Epidemiology Expert (2016-2021)

Honors: Best Young Scientist (International Society of Clinical Animal Pathology, 2006).

Publications from the Last 3 Years:

[1] Marinelli, L; Fornasari, E; Eusepi, P; Ciulla, M; Genovese, S; Epifano, F; Fiorito, S; Turkez, H; Örtücü, S; Mingoia, M; Simoni, S; Pugnaloni, A; Di Stefano, A; Cacciatore, I. Carvacrol prodrugs as novel antimicrobial agents. *Eur J Med Chem.*, 2019, Jun 10, 178, 515-529. doi: 10.1016/j.ejmech.2019.05.093. [Epub ahead of print] PubMed PMID: 31207463.

[2] Çadirci, K; Türkez, H; Özdemir, Ö. The *in vitro* cytotoxicity, genotoxicity and oxidative damage potential of the oral dipeptidyl peptidase-4 inhibitor, linagliptin, on cultured human mononuclear blood cells. *Acta Endocrinol* (Buchar)., 2019 Jan-Mar, 5(1), 9-15. doi: 10.4183/aeb.2019.9. PubMed PMID: 31149054; PubMed Central PMCID: PMC6535332.

[3] Türkez, H; Arslan, ME; Sönmez, E; Geyikoğlu, F; Açıkyıldız, M; Tatar, A. Microarray assisted toxicological investigations of boron

carbide nanoparticles on human primary alveolar epithelial cells. *Chem Biol Interact.*, 2019 Feb, 25, 300, 131-137. doi: 10.1016/j.cbi.2019.01.021. Epub 2019 Jan 24. PubMed PMID: 30684454.

[4] Türkez, H; Arslan, ME; Sönmez, E; Açikyildiz, M; Tatar, A; Geyikoğlu, F. Synthesis, characterization and cytotoxicity of boron nitride nanoparticles: emphasis on toxicogenomics. *Cytotechnology.*, 2019 Feb, 71(1), 351-361. doi: 10.1007/s10616-019-00292-8. Epub 2019 Jan 14. PubMed PMID: 30644070; PubMed Central PMCID: PMC6368500.

[5] Alak, G; Yeltekin, AÇ; Uçar, A; Parlak, V; Türkez, H; Atamanalp, M. Borax Alleviates Copper-Induced Renal Injury via Inhibiting the DNA Damage and Apoptosis in Rainbow Trout. *Biol Trace Elem Res.*, 2019 Jan, 5. doi: 10.1007/s12011-018-1622-5. [Epub ahead of print] PubMed PMID: 30612301.

[6] Çelikezen, FÇ; Hayta, Ş; Özdemir, Ö; Türkez, H. Cytotoxic and antioxidant properties of essential oil of Centaurea behen L. *in vitro*. *Cytotechnology.*, 2019 Feb, 71(1), 345-350. doi: 10.1007/s10616-018-0290-9. Epub 2019 Jan 2. PubMed PMID: 30603915; PubMed Central PMCID: PMC6368501.

[7] Marinelli, L; Fornasari, E; Di Stefano, A; Turkez, H; Genovese, S; Epifano, F; Di Biase, G; Costantini, E; D'Angelo, C; Reale, M; Cacciatore, I. Synthesis and biological evaluation of novel analogues of Gly-l-Pro-l-Glu (GPE) as neuroprotective agents. *Bioorg Med Chem Lett.*, 2019 Jan, 15, 29(2), 194-198. doi: 10.1016/j.bmcl.2018.11.057. Epub 2018 Nov 29. PubMed PMID: 30522955.

[8] Alak, G; Parlak, V; Yeltekin, AÇ; Ucar, A; Çomaklı, S; Topal, A; Atamanalp, M; Özkaraca, M; Türkez, H. The protective effect exerted by dietary borax on toxicity metabolism in rainbow trout (Oncorhynchus mykiss) tissues. *Comp Biochem Physiol C Toxicol Pharmacol.*, 2019 Feb, 216, 82-92. doi: 10.1016/j.cbpc.2018.10.005. Epub 2018 Nov 9. PubMed PMID: 30419360.

[9] Emsen, B; Aslan, A; Turkez, H; Joughi, A; Kaya, A. The anti-cancer efficacies of diffractaic, lobaric, and usnic acid: *In vitro* inhibition of

glioma. *J Cancer Res Ther.*, 2018, Jul-Sep, 14(5), 941-951. doi: 10.4103/0973-1482.177218. PubMed PMID: 30197329.

[10] Turkez, H; Tozlu, OO; Lima, TC; de Brito, AEM; de Sousa, DP. A Comparative Evaluation of the Cytotoxic and Antioxidant Activity of Mentha crispa Essential Oil, Its Major Constituent Rotundifolone, and Analogues on Human Glioblastoma. *Oxid Med Cell Longev.*, 2018, Jul 2, 2018, 2083923. doi: 10.1155/2018/2083923. eCollection 2018. PubMed PMID: 30057673; PubMed Central PMCID: PMC6051078.

[11] Alak, G; Ucar, A; Yeltekin, AÇ; Çomaklı, S; Parlak, V; Taş, IH; Özkaraca, M; Topal, A; Kirman, EM; Bolat, İ; Atamanalp, M; Türkez, H. Neuroprotective effects of dietary borax in the brain tissue of rainbow trout (Oncorhynchus mykiss) exposed to copper-induced toxicity. *Fish Physiol Biochem.*, 2018 Oct, 44(5), 1409-1420. doi: 10.1007/s10695-018-0530-0. Epub 2018 Jun 29. PubMed PMID: 29959587.

[12] Alak, G; Parlak, V; Aslan, ME; Ucar, A; Atamanalp, M; Turkez, H. Borax Supplementation Alleviates Hematotoxicity and DNA Damage in Rainbow Trout (Oncorhynchus mykiss) Exposed to Copper. *Biol Trace Elem Res.*, 2019 Feb, 187(2), 536-542. doi: 10.1007/s12011-018-1399-6. Epub 2018 Jun 21. PubMed PMID: 29926392.

[13] Koc, K; Ozdemir, O; Ozdemir, A; Dogru, U; Turkez, H. Antioxidant and anticancer activities of extract of Inula helenium (L.) in human U-87 MG glioblastoma cell line. *J Cancer Res Ther.*, 2018 Apr-Jun, 14(3), 658-661. doi: 10.4103/0973-1482.187289. PubMed PMID: 29893335.

[14] da Nóbrega, FR; Ozdemir, O; Nascimento Sousa, SCS; Barboza, JN; Turkez, H; de Sousa, DP. Piplartine Analogues and Cytotoxic Evaluation against Glioblastoma. *Molecules.*, 2018 Jun, 8, 23(6). pii: E1382. doi: 10.3390/molecules23061382. PubMed PMID: 29890617; PubMed Central PMCID: PMC6099735.

[15] Akbaba, GB; Türkez, H. Investigation of the Genotoxicity of Aluminum Oxide, β-Tricalcium Phosphate, and Zinc Oxide Nanoparticles *In Vitro. Int J Toxicol.*, 2018 May/Jun, 37(3), 216-222.

doi: 10.1177/1091581818775709. Epub 2018 May 4. PubMed PMID: 29727252.

[16] Emsen, B; Togar, B; Turkez, H; Aslan, A. Effects of two lichen acids isolated from Pseudevernia furfuracea (L.) Zopf in cultured human lymphocytes. *Z Naturforsch C.*, 2018 Jul, 26, 73(7-8), 303-312. doi: 10.1515/znc-2017-0209. PubMed PMID: 29573381.

[17] Hritcu, L; Ionita, R; Postu, PA; Gupta, GK; Turkez, H; Lima, TC; Carvalho, CUS; de Sousa, DP. Antidepressant Flavonoids and Their Relationship with Oxidative Stress. *Oxid Med Cell Longev.*, 2017, 2017, 5762172. doi: 10.1155/2017/5762172. Epub 2017 Dec 19. Review. PubMed PMID: 29410733; PubMed Central PMCID: PMC5749298.

[18] Özgeriş, B; Akbaba, Y; Özdemir, Ö; Türkez, H; Göksu, S. Synthesis and Anticancer Activity of Novel Ureas and Sulfamides Incorporating 1-Aminotetralins. *Arch Med Res.*, 2017 Aug, 48(6), 513-519. doi: 10.1016/j.arcmed.2017.12.002. Epub 2017 Dec 14. PubMed PMID: 29248174.

[19] Marinelli, L; Fornasari, E; Di Stefano, A; Turkez, H; Arslan, ME; Eusepi, P; Ciulla, M; Cacciatore, I. (R)-α-Lipoyl-Gly-l-Pro-l-Glu dimethyl ester as dual acting agent for the treatment of Alzheimer's disease. *Neuropeptides.*, 2017 Dec, 66, 52-58. doi: 10.1016/j.npep.2017.09.001. Epub 2017 Oct 2. PubMed PMID: 28993014.

[20] Geyikoglu, F; Emir, M; Colak, S; Koc, K; Turkez, H; Bakir, M; Hosseinigouzdagani, M; Cerig, S; Keles, ON; Ozek, NS. Effect of oleuropein against chemotherapy drug-induced histological changes, oxidative stress, and DNA damages in rat kidney injury. *J Food Drug Anal.*, 2017 Apr, 25(2), 447-459. doi: 10.1016/j.jfda.2016.07.002. Epub 2016 Aug 2. PubMed PMID: 28911689.

[21] Turkez, H; Arslan, ME; Ozdemir, O. Genotoxicity testing: progress and prospects for the next decade. *Expert Opin Drug Metab Toxicol.*, 2017 Oct, 13(10), 1089-1098. doi: 10.1080/17425255.2017.1375097. Epub 2017 Sep 10. Review. PubMed PMID: 28889778.

[22] Cacciatore, I; Fornasari, E; Marinelli, L; Eusepi, P; Ciulla, M; Ozdemir, O; Tatar, A; Turkez, H; Di Stefano, A. Memantine-derived

drugs as potential antitumor agents for the treatment of glioblastoma. *Eur J Pharm Sci.*, 2017 Nov, 15, 109, 402-411. doi: 10.1016/j.ejps.2017.08.030. Epub 2017 Aug 30. PubMed PMID: 28860082.

[23] Geyikoğlu, F; Çolak, S; Türkez, H; Bakır, M; Koç, K; Hosseinigouzdagani, MK; Çeriğ, S; Sönmez, M. Oleuropein Ameliorates Cisplatin-induced Hematological Damages Via Restraining Oxidative Stress and DNA Injury. *Indian J Hematol Blood Transfus.*, 2017 Sep, 33(3), 348-354. doi: 10.1007/s12288-016-0718-3. Epub 2016 Aug 16. PubMed PMID: 28824236; PubMed Central PMCID: PMC5544628.

[24] Türkez, H; Arslan, ME; Sönmez, E; Tatar, A; Açikyildiz, M; Geyikoğlu, F. Toxicogenomic responses of human alveolar epithelial cells to tungsten boride nanoparticles. *Chem Biol Interact.*, 2017, Aug 1, 273, 257-265. doi: 10.1016/j.cbi.2017.06.027. Epub 2017 Jun 27. PubMed PMID: 28666766.

[25] Carbone, C; Arena, E; Pepe, V; Prezzavento, O; Cacciatore, I; Turkez, H; Marrazzo, A; Di Stefano, A; Puglisi, G. Nanoencapsulation strategies for the delivery of novel bifunctional antioxidant/σ1 selective ligands. *Colloids Surf B Biointerfaces.*, 2017, Jul 1, 155, 238-247. doi: 10.1016/j.colsurfb.2017.04.016. Epub 2017 Apr 12. PubMed PMID: 28432957.

[26] Aydın, E; Türkez, H; Hacımüftüoğlu, F; Tatar, A; Geyikoğlu, F. Molecular genetic and biochemical responses in human airway epithelial cell cultures exposed to titanium nanoparticles *in vitro*. *J Biomed Mater Res A.*, 2017 Jul, 105(7), 2056-2064. doi: 10.1002/jbm.a.35994. Epub 2017 May 17. PubMed PMID: 28028929.

[27] Crescenzo, AD; Cacciatore, I; Petrini, M; D'Alessandro, M; Petragnani, N; Boccio, PD; Profio, PD; Boncompagni, S; Spoto, G; Turkez, H; Ballerini, P; Stefano, AD; Fontana, A. Gold nanoparticles as scaffolds for poor water soluble and difficult to vehiculate antiparkinson codrugs. *Nanotechnology.*, 2017, Jan 13, 28(2), 025102. Epub 2016 Dec 6. PubMed PMID: 27922827.

[28] Aydın, E; Turkez, H; Tasdemir, S; Hacımuftuoglu, F. Anticancer, Antioxidant and Cytotoxic Potential of Thymol *in vitro* Brain Tumor

Cell Model. *Cent Nerv Syst Agents Med Chem.*, 2017, 17(2), 116-122. doi: 10.2174/1871524916666160823121854. PubMed PMID: 27554922.

[29] Cacciatore, I; Marinelli, L; Fornasari, E; Cerasa, LS; Eusepi, P; Türkez, H; Pomilio, C; Reale, M; D'Angelo, C; Costantini, E; Di Stefano, A. Novel NSAID-Derived Drugs for the Potential Treatment of Alzheimer's Disease. *Int J Mol Sci.*, 2016, Jun 30, 17(7). pii: E1035. doi: 10.3390/ijms17071035. PubMed PMID: 27376271; PubMed Central PMCID: PMC4964411.

[30] Fornasari, E; Marinelli, L; Di Stefano, A; Eusepi, P; Turkez, H; Fulle, S; Di Filippo, ES; Scarabeo, A; Di Nicola, S; Cacciatore, I. Synthesis and Antioxidant Properties of Novel Memantine Derivatives. *Cent Nerv Syst Agents Med Chem.*, 2017, 17(2), 123-128. doi: 10.2174/1871524916666160625123621. PubMed PMID: 27356627.

[31] Arena, E; Cacciatore, I; Cerasa, LS; Turkez, H; Pittalà, V; Pasquinucci, L; Marrazzo, A; Parenti, C; Di Stefano, A; Prezzavento, O. New bifunctional antioxidant/σ1 agonist ligands: Preliminary chemico-physical and biological evaluation. *Bioorg Med Chem.*, 2016, Jul 15, 24(14), 3149-56. doi: 10.1016/j.bmc.2016.05.045. Epub 2016 May 21. PubMed PMID: 27262426.

[32] Emsen, B; Turkez, H; Togar, B; Aslan, A. Evaluation of antioxidant and cytotoxic effects of olivetoric and physodic acid in cultured human amnion fibroblasts. *Hum Exp Toxicol.*, 2017 Apr, 36(4), 376-385. doi: 10.1177/0960327116650012. Epub 2016 May 20. PubMed PMID: 27206701.

[33] Tasdemir, S; Eroz, R; Dogan, H; Erdem, HB; Sahin, I; Kara, M; Engin, RI; Turkez, H. Association Between Human Hair Loss and the Expression Levels of Nucleolin, Nucleophosmin, and UBTF Genes. *Genet Test Mol Biomarkers.*, 2016 Apr, 20(4), 197-202. doi: 10.1089/gtmb.2015.0246. Epub 2016 Feb 11. PubMed PMID: 26866305.

[34] Cacciatore, I; Fornasari, E; Di Stefano, A; Marinelli, L; Cerasa, LS; Turkez, H; Aydin, E; Moretto, A; Ferrone, A; Pesce, M; di Giacomo, V; Reale, M; Costantini, E; Di Giovanni, P; Speranza, L; Felaco, M:

Patruno, A. Development of glycine-α-methyl-proline-containing tripeptides with neuroprotective properties. *Eur J Med Chem.*, 2016 Jan 27, 108, 553-563. doi: 10.1016/j.ejmech.2015.12.003. Epub 2015 Dec 10. PubMed PMID: 26717205.

[35] Emsen, B; Aslan, A; Togar, B; Turkez, H. *In vitro* antitumor activities of the lichen compounds olivetoric, physodic and psoromic acid in rat neuron and glioblastoma cells. *Pharm Biol.*, 2016 Sep, 54(9), 1748-62. doi: 10.3109/13880209.2015.1126620. Epub 2015 Dec 24. PubMed PMID: 26704132.

[36] Sonmez, E; Cacciatore, I; Bakan, F; Turkez, H; Mohtar, YI; Togar, B; Stefano, AD. Toxicity assessment of hydroxyapatite nanoparticles in rat liver cell model *in vitro*. *Hum Exp Toxicol.*, 2016 Oct, 35(10), 1073-83. doi: 10.1177/0960327115619770. Epub 2015 Dec 11. PubMed PMID: 26655636.

[37] Hacimuftuoglu, A; Tatar, A; Cetin, D; Taspinar, N; Saruhan, F; Okkay, U; Turkez, H; Unal, D; Stephens, RL; Jr. Suleyman, H. Astrocyte/neuron ratio and its importance on glutamate toxicity: an in vitro voltammetric study. *Cytotechnology.*, 2016 Aug, 68(4), 1425-33. doi: 10.1007/s10616-015-9902-9. Epub 2015 Oct 6. PubMed PMID: 26438331; PubMed Central PMCID: PMC4960189.

[38] Bakır, M; Geyikoglu, F; Colak, S; Turkez, H; Bakır, TO; Hosseinigouzdagani, M. The carvacrol ameliorates acute pancreatitis-induced liver injury via antioxidant response. *Cytotechnology.*, 2016 Aug, 68(4), 1131-46. doi: 10.1007/s10616-015-9871-z. Epub 2015 Sep 8. PubMed PMID: 26350272; PubMed Central PMCID: PMC4960162.

[39] Cetin, D; Hacımuftuoglu, A; Tatar, A; Turkez, H; Togar, B. The *in vitro* protective effect of salicylic acid against paclitaxel and cisplatin-induced neurotoxicity. *Cytotechnology.*, 2016 Aug, 68(4), 1361-7. doi: 10.1007/s10616-015-9896-3. Epub 2015 Jul 22. PubMed PMID: 26199062; PubMed Central PMCID: PMC4960183.

[40] Kılıç, Y; Geyikoglu, F; Çolak, S; Turkez, H; Bakır, M; Hsseinigouzdagani, M. Carvacrol modulates oxidative stress and decreases cell injury in pancreas of rats with acute pancreatitis.

Cytotechnology., 2016 Aug, 68(4), 1243-56. doi: 10.1007/s10616-015-9885-6. Epub 2015 Jun 21. PubMed PMID: 26093481; PubMed Central PMCID: PMC4960173.

[41] Çelikezen, FÇ; Toğar, B; Özgeriş, FB; İzgi, MS; Türkez, H. Cytogenetic and oxidative alterations after exposure of cultured human whole blood cells to lithium metaborate dehydrate. *Cytotechnology.*, 2016 Aug, 68(4), 821-7. doi: 10.1007/s10616-014-9833-x. Epub 2015 Feb 14. PubMed PMID: 25680697; PubMed Central PMCID: PMC4960131.

[42] Çolak, S; Geyikoğlu, F; Bakır, TÖ; Türkez, H; Aslan, A. Evaluating the toxic and beneficial effects of lichen extracts in normal and diabetic rats. *Toxicol Ind Health.*, 2016 Aug, 32(8), 1495-1504. Epub 2015 Jan 29. PubMed PMID: 25647809.

[43] Akbaba, GB; Turkez, H; Sönmez, E; Tatar, A; Yilmaz, M. Genotoxicity in primary human peripheral lymphocytes after exposure to lithium titanate nanoparticles *in vitro*. *Toxicol Ind Health.*, 2016 Aug, 32(8), 1423-1429. Epub 2014 Dec 31. PubMed PMID: 25552539.

[44] Turkez, H; Geyikoglu, F; Yousef, MI. Ameliorative effects of docosahexaenoic acid on the toxicity induced by 2,3,7,8-tetrachlorodibenzo-p-dioxin in cultured rat hepatocytes. *Toxicol Ind Health.*, 2016 Jun, 32(6), 1074-85. doi: 10.1177/0748233714547382. Epub 2014 Sep 3. PubMed PMID: 25187318.

[45] Turkez, H; Sönmez, E; Di Stefano, A; Mokhtar, YI. Health risk assessments of lithium titanate nanoparticles in rat liver cell model for its safe applications in nanopharmacology and nanomedicine. *Cytotechnology.*, 2016 Mar, 68(2), 291-302. doi: 10.1007/s10616-014-9780-6. Epub 2014 Aug 23. PubMed PMID: 25149287; PubMed Central PMCID: PMC4754244.

[46] Türkez, H; Aydın, E. Investigation of cytotoxic, genotoxic and oxidative properties of carvacrol in human blood cells. *Toxicol Ind Health.*, 2016 Apr, 32(4), 625-33. doi: 10.1177/0748233713506771. Epub 2013 Nov 8. PubMed PMID: 24215060.

[47] Deniz, GY; Geyikoğlu, F; Türkez, H; Bakır, TÖ; Çolak, S; Aslan, A. The biochemical and histological effects of lichens in normal and diabetic rats. *Toxicol Ind Health.*, 2016 Apr, 32(4), 601-13. doi: 10.1177/0748233713506769. Epub 2013 Nov 5. PubMed PMID: 24193057.

[48] Polat, Z; Aydın, E; Türkez, H; Aslan, A. *In vitro* risk assessment of usnic acid. *Toxicol Ind Health.*, 2016 Mar, 32(3), 468-75. doi: 10.1177/0748233713504811. Epub 2013 Nov 5. PubMed PMID: 24193043.

[49] Türkez, H; Aydın, E. *In vitro* assessment of cytogenetic and oxidative effects of α-pinene. *Toxicol Ind Health.*, 2016 Jan, 32(1), 168-76. doi: 10.1177/0748233713498456. Epub 2013 Sep 30. PubMed PMID: 24081629.

In: Micronucleus Assay: An Overview
Editor: Robert C. Cole

ISBN: 978-1-53616-678-1
© 2020 Nova Science Publishers, Inc.

Chapter 4

MICRONUCLEUS ASSAY IN OCCUPATIONAL TOXICOLOGY STUDIES

*Hatice Gül Anlar**

Department of Pharmaceutical Toxicology, Faculty of Pharmacy, Zonguldak Bulent Ecevit University, Zonguldak, Turkey

ABSTRACT

The micronucleus (MN) was recognized at the end of the 19th century when Howell and Jolly found small inclusions in the blood taken from cats and rats. The small inclusions, called Howell-Jolly body, are also observed in the erythrocytes of peripheral blood from severe anemia patients. After that, the MN test is widely used in toxicological studies and now recognized as one of the most successful and reliable assays for genotoxic carcinogens, i.e., carcinogens that act by causing genetic damage. There are two major versions of this test i.e., *in vivo* and *in vitro*. This test also widely used in occupational toxicology studies. A person spends, on average, one/third of his/her life at his/her workplace and therefore the environment in which he/she works can be a major factor in determining health status. Many studies have confirmed that the number of MN have increased in workers exposed to inorganic lead, painters exposed to lead-

---

* Corresponding Author's Email: haticegulanlar@gmail.com.

containing pigments, ceramic dust, polycyclic aromatic hydrocarbons, coal dust, and welding fume. MN can be evaluated in different kinds of cells that do not necessarily have to divide *in vitro* such as epithelial cells, thus, the analysis of MN in exfoliated buccal cells has been demonstrated to be a sensitive method for monitoring genetic damage in human populations. Since epithelial cells can be obtained easily by relatively non-invasive methods and are capable to indicate toxicity in actual target tissue by the MN assay, their usage in case-control studies has been increasing. This chapter aims to provide knowledge about MN assay and its usage in occupational toxicology studies.

Keywords: micronucleus assay, occupational toxicology, workers, health, genotoxicity

INTRODUCTION

The evaluation of micronuclei (MN) has been focused by researchers in order to determine the toxic effects of genotoxic exposure. In comparison with other genotoxicity assays like chromosomal aberrations (CA), MN assay is simpler due to its simpler scoring and training. Also, MN assay is more sensitive than the CA assay, because thousands of cells are counted in MN assay while only a hundred or a few hundred cells are usually scored for CA (Norppa and Falck 2003). In addition to that, the ability of the MN assay to detect both clastogenic and aneugenic effects (leading to structural and numerical chromosome alterations, respectively) is an advantage of the MN technique. There are two main versions of MN assay in human studies i.e., cytokinesis block MN assay (CBMN) and MN assay in exfoliated cells. In humans, most MN studies have been conducted using cultured peripheral lymphocytes, which lend themselves well to both genotoxicity testing and biomonitoring. The CBMN, based on cytokinesis inhibition by cytochalasin B (Cyt B), has facilitated MN analysis exclusively in binucleate cells that have completed their first *in vitro* division after treatment with the test compound or following culture initiation (Fenech et al. 1999). Recently, the analysis of MN in exfoliated cells such as buccal and nasal cells has been increasingly used in occupational biomonitoring studies since these cells can be obtained easily by relatively non-invasive methods and are capable to indicate toxicity in the actual target tissue (Sarto et al. 1987).

In order to evaluate the health status of a person, it should be considered that her/his workplace because a person spends, on average, one-third of her/his life at the workplace. Therefore determination of toxic effects due to the occupational exposure is really important, especially in the developing countries. MN assay, both two forms, has been used in occupational studies for example with workers exposed to inorganic lead, painters exposed to lead-containing pigments, ceramic dust, polycyclic aromatic hydrocarbons (PAHs), coal dust, and welding fume (Palanikumar and Panneerselvam 2011). MN is a reliable biomarker for the detection of increased human cancer risks (Bonassi, El-Zein, et al. 2011, Bonassi et al. 2007).

This chapter aims to provide knowledge about MN assay and its usage in occupational toxicology studies with different occupational settings.

HISTORY OF MN ASSAY

The history of the MN assay began in 1971 when Swiss cytogeneticists Matter and Schmid who worked at the Department of Paediatrics in the University of Zürich detected MN in extracellular DNA containing bodies in bone marrow cells of mice and hamsters (Miller 1973). After, it was understood that MN is formed as a consequence of structural and numerical chromosomal aberrations (CA) and can be scored much faster than these structures in metaphase cells. Chromosome breakage and dysfunction of the mitotic apparatus are leading to the formation of MN in mitotic cells. MN is formed from acentric chromosome or chromatid fragments and whole chromosomes or chromatids that lag behind in anaphase and are left outside the nuclei in telophase (Lindholm et al. 1991).

In the following years, a variety of tests was developed which are based on the quantification of MN (Heddle et al. 2011). But only a few publications which concerned the formation of MN in occupationally exposed workers were published before the protocol for the lymphocyte CBMN assay using cytochalasin-B was developed in 1985. The development of this protocol was an important milestone because it was clear from *in vitro* studies with genotoxins that the frequency of MN varies depending on the proportion of dividing cells (Fenech and Morley 1985a). To overcome this kinetic problem, Fenech and Morley (Fenech and Morley 1985b) in 1985, developed a method to identify cells that have completed nuclear division by their binucleated appearance by using cytochalasin-B, an inhibitor of cytokinesis. The development of the CBMN method made the assay much more robust by eliminating false-negative results caused by inhibition of nuclear division by the agent being tested or as a result of host factors (such as aging) that reduce lymphocyte responsiveness to mitogen. In addition, the CBMN assay allowed the proportion of dividing cells to be measured and it became possible to score nucleoplasmic bridges (a biomarker of DNA strand break misrepair or telomere end-fusions) in binucleated cells which enabled a more comprehensive assessment of genotoxicity. But, there were other problems in the use of the lymphocyte CBMN assay such as different fixation techniques, staining procedures, scoring methods and scoring criteria i.e., the number of binucleated cells scored per sample (Bonassi et al. 2001). In order to reduce this heterogeneity, which led to substantial intra- and inter-laboratory variations, the HUMN (HUman MicroNucleus) consortium was formed in 1997 (Fenech et al. 1999). This consortium published a standard protocol of MN assay which includes results of scoring exercises, main confounding factors as well as results of statistical analyses (Fenech et al. 2011).

Fluorescence *in situ* hybridization technique (FISH) was another important achievement in which "pancentromeric probes" were used to make a distinction between MN which contain chromosomal fragments and MN with centromeres which contain entire chromosomes (Doherty et al. 1996). Later on, automated scoring devices which enable to evaluate larger sample sizes and to avoid inter-individual scorer variations were developed but only used in approximately 20 studies since it is a very expensive system (Fenech et al. 2013).

In 2006 the "cytome" assay version of the lymphocyte CBMN method for measuring comprehensively chromosomal instability phenotypes and altered cellular viability caused by genetic defects, nutritional deficiencies and exogenous exposure to genotoxins was introduced (Fenech 2006). In this assay, nuclear anomalies apart from MN such as nucleoplasmic bridges (formed as a consequence of dicentric chromosome formation) and nuclear buds (reflecting expulsion of amplified DNA or unresolved DNA repair complexes) can be analyzed (Fenech et al. 2011). Additionally, counting these anomalies provide information on acute cytotoxic effects i.e., apoptosis and necrosis. The nuclear division index (NDI) is a measure of the proliferative status of the viable cell fraction, therefore, it is an indicator of cytostatic effects and, in the case of lymphocytes, also a measure of mitogenic response which is a biomarker of immune functions (Fenech et al. 2003). MN assay is widely used for ecotoxicology studies, determination of radiation sensitivity during cancer risk assessment, regulation of radiation treatment, and investigation of cytogenetic properties of new drugs and chemicals (Kirsch-Volders et al. 1997).

The *in vitro* MN test with lymphocytes and the *in vivo* bone marrow MN assay in rodents are currently included in the Organisation for Economic Co-operation and Development (OECD) guidelines for screening of chemicals (OECD, 2016).

THE MN ASSAY IN EXFOLIATED CELL

Studies with MN assay in exfoliated cells such as buccal and nasal epithelial cells has greatly increased in the last decade (Holland et al. 2008). Because exfoliated cells are rapidly dividing and allow the assessment of DNA damage without the need for an ex vivo cell replication step. Also, cell collection is non-invasive and easy. Therefore this assay is well suited for large biomonitoring studies, especially in pediatric populations. Other advantages are the specificity for detecting the effects of exposure to inhaled or ingested genotoxic agents. In addition, MN frequency in exfoliated cells is correlated with MN frequency in lymphocytes (Ceppi et al. 2010).

Epithelial cells are rapidly proliferating tissues that consist of cellular and vascular layers that surround all body surfaces, cavities and glands and are in constant contact with the environment. Buccal and nasal epithelial cells can be easily collected with small brushes, wooden sticks or cotton-tipped swabs. It is also possible to take samples as often as desired. But it is more difficult to take samples from the esophagus and bronchial tissues and it takes some time to get the samples again. Exfoliated epithelial cells can also be obtained from urine by centrifugation. Since exfoliated cells are not viable, the MN evaluation in these cells should be performed at the interphase stage (Moore et al. 1996).

Before performing the MN assay, it is necessary to know the kinetics of MN formation in epithelial cells. These cells are classified according to the number of layers of cells in the tissue and the shape of the superficial cells. For example, there is multilayer squamous epithelium in the cervix and oral cavity. These tissues are composed of basal, intermediate and superficial cells. The basal layer consists of self-regenerating stem cells. These cells move towards the surface to form sister cells. The stratum corneum is the keratinized layer and below this layer, there are stratum granulosa and stratum spinosum. In the stratum germinativum which is at the bottom, there are basal cells. The time required for the cell division depends on the cell type but is generally about 25 days. Also, irritation caused by physical, chemical or infectious factors may affect the rate of regeneration of tissues. MN occurs as a result of chromosomal damage in the basal cells of the epithelium. After the cell division, this structure which is surrounded by membranes in the cytoplasm is discarded. MN assay in exfoliated cells has more advantageous than the MN test in lymphocytes when the target tissue is epithelial tissue. The target tissue can be examined directly with MN assay in exfoliated cells. In addition, the relationship between short-term cytogenetic effects and long-term carcinogenic effects can be investigated much better in epithelial cells than lymphocytes (Thomas et al. 2009).

The new project HUMNXL ('XL' referring to eXfoLiated cells) was designed for the determination of technical, protocol and life-style variables affecting MN frequency and use this information to design intra- and inter-laboratory validation studies of the method. Another aim of this project is the evaluation of the role of buccal MN and other nuclear alterations in the prediction of cancer and other degenerative diseases (Bonassi, Coskun, et al. 2011).

BUCCAL MN CYTOME ASSAY (BMCYT)

Buccal epithelial cells are the first barrier during the entry of chemicals into the body by inhalation or orally, and are capable of metabolizing carcinogenic substances into active products (Autrup et al. 1985, Liu et al. 1993, Vondracek et al. 2001, Spivack et al. 2004). Buccal epithelial cells are the most used exfoliated cell in MN assay. Since up to 90% of all cancers appear to be the epithelial origin, these cells are a suitable model for biomonitoring (Rosin 1992). It is a very simple and painless procedure compared to the collection of blood and tissue biopsies. Therefore, BMCyt has become a widely used method for monitoring genotoxic damage worldwide.

In this method, cells are taken from the inner cheeks of the subjects with a wooden spatula, small-headed toothbrush or metal spatula. Cells should be taken from both sides of cheeks. Simultaneous repetitive sampling may result in the collection of less differentiated cells. For the assessment of acute exposure, it is ideal to take at least one sample every seven days and repeat the sampling after 28 days or later. In the case of chronic exposure, sampling is recommended at least once every 3 months. After sampling, alternative methods such as methanol: glacial acetic acid (3: 1) or 80% methanol may be used for the fixation of the cells. There are also different staining methods. However, the use of DNA-specific dyes such as Feulgen and Acridine Orange should be preferred to reduce the false-positive results which are commonly seen in Romanowsky dyes such as Giemsa and Leishmann's. The frequencies of MN in buccal epithelial cells in smokers and nonsmokers were investigated with different staining techniques. When the Romanowsky dyes were used, the frequencies of MN in smokers were found to be 4-5 times higher than non-smokers. However, when DNA-specific dyes were used, no significant difference was found between the MN frequencies. When Feulgen dye was used DNA material is seen with bright red color under red filtered fluorescent light (580-620 nm emission wavelength) and thus false-positive results can be prevented in the evaluation (Casartelli et al. 1997, Nersesyan et al. 2006, Thomas et al. 2009).

There are several endpoints can be evaluated this assay i.e., MN, nuclear bud (NB), binucleated cells (BN), condensed chromatin (CC), karyorrhectic (KH), pyknotic (PC) or karyolitic (KL) cells as well as the frequency of basal and fully differentiated cells. The counting of the MN in the buccal epithelial cells is based on the criteria defined by Tolbert et al. (Tolbert, Shy, and Allen 1992). In these criteria; cells are divided into two categories i.e., normal and abnormal cells. Normal basal cells are smaller and oval compared to differentiated cells and have a larger nucleus/cytoplasm ratio. The shape of the differentiated cells is more angular and flatter. In normal basal and differentiated cells, there are no structures containing DNA outside the main nucleus. MN is oval or round and 1/3 to 1/16 of the nucleus. MNs are counted in differentiated cells but it may also be seen in basal cells but it is not practical to evaluate because of its low frequency. NB is a thinner and distinct structure located at the end of the main nucleus and is also called broken egg. NB and the main nucleus are connected to each other. The mechanism of NB formation is not fully understood but is associated with errors in DNA repair or loss of amplified DNA (Thomas et al. 2009). BN cells contain two main nuclei within a cell. The importance of these cells is unknown but is thought to indicate cytokinesis errors (Bonassi et al. 2001). In CC cells, chromatins were gathered in some regions of the nucleus while those in some regions are disappeared. Nuclear disintegration may also occur in these cells and which may indicate an early stage of apoptosis. KH cells have more concentrated chromatins than CC cells. These cells may indicate advanced stages of apoptosis. PC cells have small and contracted nuclei. The mechanism leading to the formation of PC cells and the biological significance of these cells is not fully understood. KL cells show the advanced stage of cell death and their nuclei are completely lost. When stained with Feulgen, the shadow of the nucleus is seen and the nucleus DNA is completely depleted (Thomas et al. 2009).

It was shown that the number of MNs in buccal epithelial cells increased 16-fold after radiotherapy in oral cancer patients (Moore et al. 1996). This method is also used to diagnosis of various diseases such as Bloom's syndrome, Down syndrome and Alzheimer's disease which is indicate the diagnostic value of the BMCyt assay (Rosin and German 1985, Thomas et al. 2008).

OCCUPATIONAL TOXICOLOGY STUDIES WITH MN ASSAY

In 2004, the first occupational study was published which concerned polymorphisms of genes encoding for gluthatione-S-transferases and other enzymes related to the formation of MN in peripheral lymphocytes of styrene exposed workers (Teixeira et al. 2004). In our laboratory, this assay has been widely used for a long time in different occupational groups. Also, we did MN assay in animals and cell cultures to evaluate the genotoxic and anti-genotoxic effects of some phenolic compounds. A first occupational study in our laboratory using MN assay was conducted with buccal epithelial cells of ceramic workers (n = 99) and their controls (n = 81). Ceramic workers exposed to a complex mixture of chemicals that can be associated with an increased risk of several diseases. The results of this study showed that buccal MN frequency of the workers was found to be significantly higher than the control group ($p < 0.05$). Workers working in the ceramic plant for more than 16 years and using alcohol had significantly higher buccal MN frequencies ($p < 0.05$). But there was no significant correlation between smoking, age, using protective equipment and buccal MN frequency. Workers with silicosis and suspected to have silicosis had higher buccal MN frequencies compared to other workers, but the differences were not statistically significant. When evaluating the frequencies of abnormal cells other than MN, it was shown that CC+KL and PYC cell frequencies were significantly higher than the control group ($p < 0.05$) (Anlar et al. 2018).

In another occupational study of our research group, we investigated buccal MN frequencies of welders who are exposed to a number of hazardous compounds such as ultraviolet (UV) radiation, electromagnetic fields, toxic metals, PAHs and welding fumes. In this study, 48 welders and an equal number of control subjects were evaluated for DNA damage in their buccal cells using BMCyt assay. Buccal MN frequencies of the workers were found to be significantly higher than the control group ($p < 0.05$). There was no significant correlation between buccal MN frequency and smoking, alcohol consumption, duration of exposure and age. When evaluating the frequencies of different cells other than MN in buccal mucosa cells, it was observed that BN, CC, KH, and KL cell frequencies in welders were significantly higher than those in the control group ($p < 0.05$). The frequencies of PYC and NB were also higher in the workers compared to controls. However, there was no statistically significant difference between the groups (Aksu et al. 2018).

We did another occupational study with boron workers who worked in boron mines in Turkey. 102 male workers who were occupationally exposed to boron from Bandırma and 110 workers who were occupationally and environmentally exposed to boron from Bigadic participated in this study. No significant difference was found in the MN frequencies between the low, medium, high and over-exposure groups. Buccal MN frequency observed in the very low exposure group was significantly lower than the all exposed groups ($p = 0.042$) but the sample size of the very low exposure group was very small. On the other hand, when evaluating the abnormal cells other than MN in buccal mucosa cells, no statistically difference between the groups was found. No correlation was seen between blood boron levels and frequencies of MN, BN, KH, KL, PYC, and NB in the buccal cell samples. A negative correlation was found between blood boron levels and frequencies of CC (Başaran et al. 2018).

PAHs are ubiquitous in our environment as they may be formed during any incomplete combustion of organic material. Engine exhaust and used engine oils are major PAH sources in engine repair workshops and traffic. It was shown that the buccal MN frequencies of exposed traffic police and taxi drivers were higher than their respective control group. Smokers and nonsmokers do not differ with respect to the incidence of MN in all groups (Karahalil, Karakaya, and Burgaz 1999).

Pesticides are chemical compounds that are used to kill pests, including insects, rodents, fungi, and unwanted plants. By their nature, pesticides are potentially toxic to other organisms, including humans, and need to be used safely and disposed of properly. Occupational pesticides exposure is also a really important issue. In order to determine the genotoxic effects of occupational pesticides exposure, MN assay has been widely used. In a study from India with 50 pesticide workers and 50 controls, MN and other nuclear abnormalities frequencies in buccal cells of the exposed group were significantly higher than those in control groups ($p < 0.05$) and also significantly related to smoking, tobacco chewing, and alcohol drinking habit ($p < 0.05$) (RafiqKhan et al. 2014). Similarly, female farmers and children from Mexico had higher buccal MN and nuclear abnormalities frequencies (Castañeda et al. 2016). In addition to that, the sensitivity and reliability of the CBMN assay have been shown to be an effective tool to measure cytogenetic damage by pesticides in several populations (Bolognesi et al. 1993, da Silva Augusto et al. 1997, Joksic, Vidakovic, and Spasojevic-Tisma 1997, Falck et al. 1999).

The painting was classified as an occupation that increases the risk of certain cancers by International Agency for Research on Cancer (IARC). Painters (n = 60) had increased buccal MN frequency compared with the control group (n = 60) ($p < 0.05$). Also, repair index (RI) was less in painters than in the controls (Celik, Diler, and Eke 2010).

Open-cast coal mine workers are potentially exposed to coal dust including quartz, trace metals, inorganic minerals, and PAHs. MN assay has been widely used for the genotoxic evaluation of occupational coal dust exposure. In a study by Donbak et al. (Donbak et al. 2005), coal workers (n = 39) had higher MN frequency in lymphocytes compared to controls. Ulker et al. (Ulker et al. 2008) investigated MN frequencies in lymphocytes of coal workers (n = 29) and coal workers pneumoconiosis patients (CWP, n = 23). MN frequencies were significantly higher in CWP patients than in coal worker and controls. On the contrary, they didn't find any differences in MN frequencies between coal worker and controls. Also, no correlation between MN frequency, smoking, duration of exposure, and age was observed in all groups. In a study from Colombia, BMCyt assay was used to detect cytogenetic damage in 100 coal workers. MN, NB, KH, and KL cell frequencies were significantly higher in the exposed group. No significant correlation between age, alcohol consumption, duration of exposure and MN assay was found in this study (Leon-Mejia et al. 2014). MN and BN cell frequencies were also found to be significantly higher in Brazilian coal workers (n = 41) (Rohr et al. 2013). Another study from Brazil showed that coal workers (n = 158) had significantly higher buccal MN frequency compared to the control group. But they didn't any significant difference between the workers and controls by CBMN assay (da Silva Junior et al. 2018).

Many studies have also confirmed that the number of MN have increased in occupational workers exposed to inorganic lead(Vaglenov, Carbonell, and Marcos 1998, Vaglenov et al. 2001), painters exposed to lead-containing pigments(Pinto et al. 2000). Rozgaj et al. (Rozgaj, Kasuba, and Jazbec 2001) have reported that MN assay is the most sensitive indicator of changes caused by anesthetic gases in hospital workers. Chang et al. (Chang et al. 1996) reported increased MN formation in nurses occupationally exposed to anesthetic gas N_2O. Pastor et al. (Pastor et al. 2003) have reported that MN analysis is the only biomarker that allows the evaluation of both clastogenic and aneuploidogenic effects in a vast range of cells since they are detected in interphase.

Jai Dong Moon et al. (Moon et al. 1998) reported that there was an increase in the frequency of MN in lymphocytes of petrochemical workers. Bolognesi et al. (Bolognesi, Perrone, and Landini 2002) reported that MN frequency in peripheral blood lymphocytes has been increased in workers involved in ornamental crop production. On the other hand, Huvinen et al. (Huvinen et al. 2002) reported that stainless steel production workers, with low exposure to dust or fumes containing hexavalent or trivalent chromium, have shown no increase in nasal MN.

CONCLUSION

Exposure to chemical carcinogens in occupational settings is usually more often than environmental exposure of the general population. Cancer rates are higher in workers exposed to multiple complexes of chemicals in many industries. Therefore monitored for any long term adverse effects of the hazardous exposure of workers is extremely important. In this chapter, information about MN assay, both in lymphocytes and exfoliated cells, and occupational toxicology studies with this assay in different occupational settings were given. Similar to other occupational studies with other genotoxicity assays, the most important aim of occupational studies was to the determination of the differences between exposed and control groups and the impact of lifestyle factors on genotoxicity. But the results of these studies regarding the lifestyle variables, especially the effect of cigarette smoking on MN frequency was contradictory. Therefore, exposure time, a number of cigarettes smoked per day, the kind of tobacco smoked as a confounding factor should be well considered.

The MN assay has the potential to be applied to all of the categories of human biomonitoring and represents a valuable tool for acquiring knowledge about current levels of exposure to occupational hazards. Especially, BMCyt assay is a useful and minimally invasive method for monitoring genetic damage in humans. But there is a need for future interlaboratory studies to widen the usage of MN assay and to eliminate its disadvantages.

REFERENCES

Aksu, İ., Anlar, H. G., Taner, G., Bacanlı, M., İritaş, S., Tutkun, E. & Basaran, N. (2018). "Assessment of DNA damage in welders using comet and micronucleus assays." *Mutation Research/Genetic Toxicology and Environmental Mutagenesis.* https://doi.org/https://doi.org/10.1016/j.mrgentox.2018.11.006.

Anlar, H. G., Taner, G., Bacanli, M., Iritas, S., Kurt, T., Tutkun, E., Yilmaz, O. H. & Basaran, N. (2018). "Assessment of DNA damage in ceramic workers." *Mutagenesis*, *33* (1), 97-104. doi: 10.1093/mutage/ gex016.

Autrup, H., Seremet, T., Arenholt, D., Dragsted, L. & Jepsen, A. (1985). "Metabolism of benzo [a] pyrene by cultured rat and human buccal mucosa cells." *Carcinogenesis*, *6* (12), 1761-1765.

Başaran, N. Y. Duydu., Üstündağ, A., Taner, G., Aydin, S., Anlar, H. G., Yalçın, C. Ö., Bacanli, M., Aydos, K., Atabekoğlu, C. S., Golka, K.,

Ickstadt, K., Schwerdtle, T., Werner, M., Meyer, S. & Bolt, H. M. (2018). "Evaluation of the DNA damage in lymphocytes, sperm and buccal cells of workers under environmental and occupational boron exposure conditions." *Mutation Research/Genetic Toxicology and Environmental Mutagenesis.* https://doi.org/https://doi.org/10.1016/j.mrgentox.2018.12.013.

Bolognesi, C., Parrini, M., Merlo, F. & Bonassi, S. (1993). "Frequency of micronuclei in lymphocytes from a group of floriculturists exposed to pesticides." *J Toxicol Environ Health, 40* (2-3), 405-11. doi: 10.1080/15287399309531807.

Bolognesi, C., Perrone, E. & Landini, E. (2002). "Micronucleus monitoring of a floriculturist population from western Liguria, Italy." *Mutagenesis, 17* (5), 391-397.

Bonassi, S., Coskun, E., Ceppi, M., Lando, C., Bolognesi, C., Burgaz, S., Holland, N., Kirsh-Volders, M., Knasmueller, S., Zeiger, E., Carnesoltas, D., Cavallo, D., da Silva, J., de Andrade, V. M., Demircigil, G. C., Dominguez Odio, A., Donmez-Altuntas, H., Gattas, G., Giri, A., Giri, S., Gomez-Meda, B., Gomez-Arroyo, S., Hadjidekova, V., Haveric, A., Kamboj, M., Kurteshi, K., Martino-Roth, M. G., Montero Montoya, R., Nersesyan, A., Pastor-Benito, S., Favero Salvadori, D. M., Shaposhnikova, A., Stopper, H., Thomas, P., Torres-Bugarin, O., Yadav, A. S., Zuniga Gonzalez, G. & Fenech, M. (2011). "The HUman MicroNucleus project on eXfoLiated buccal cells (HUMN(XL)): the role of life-style, host factors, occupational exposures, health status, and assay protocol." *Mutat Res, 728* (3), 88-97. doi: 10.1016/j.mrrev.2011.06.005.

Bonassi, S., El-Zein, R., Bolognesi, C. & Fenech, M. (2011). "Micronuclei frequency in peripheral blood lymphocytes and cancer risk: evidence from human studies." *Mutagenesis, 26* (1), 93-100. doi: 10.1093/mutage/geq075.

Bonassi, S., Fenech, M., Lando, C., Lin, Y. P., Ceppi, M., Chang, W. P., Holland, N., Kirsch-Volders, M., Zeiger, E., Ban, S., Barale, R., Bigatti, M. P., Bolognesi, C., Jia, C., Di Giorgio, M., Ferguson, L. R., Fucic, A., Lima, O. G., Hrelia, P., Krishnaja, A. P., Lee, T. K., Migliore, L.,

Mikhalevich, L., Mirkova, E., Mosesso, P., Muller, W. U., Odagiri, Y., Scarffi, M. R., Szabova, E., Vorobtsova, I., Vral, A. & Zijno, A. (2001). "HUman MicroNucleus project: international database comparison for results with the cytokinesis-block micronucleus assay in human lymphocytes: I. Effect of laboratory protocol, scoring criteria, and host factors on the frequency of micronuclei." *Environ Mol Mutagen*, *37* (1), 31-45.

Bonassi, S., Znaor, A., Ceppi, M., Lando, C., Chang, W. P., Holland, N., Kirsch-Volders, M., Zeiger, E., Ban, S., Barale, R., Bigatti, M. P., Bolognesi, C., Cebulska-Wasilewska, A., Fabianova, E., Fucic, A., Hagmar, L., Joksic, G., Martelli, A., Migliore, L., Mirkova, E., Scarfi, M. R., Zijno, A., Norppa, H. & Fenech, M. (2007). "An increased micronucleus frequency in peripheral blood lymphocytes predicts the risk of cancer in humans." *Carcinogenesis*, *28* (3), 625-31. doi: 10.1093/carcin/bgl177.

Casartelli, G., Monteghirfo, S., De Ferrari, M., Bonatti, S., Scala, M., Toma, S., Margarino, G. & Abbondandolo, A. (1997). "Staining of micronuclei in squamous epithelial cells of human oral mucosa." *Anal Quant Cytol Histol*, *19* (6), 475-81.

Castañeda, I., Arellano, E., Garcia-Zarate, M., Balam, R., Zavala-Cerna, M. & Bugarin, O. (2016). "Biomonitoring with Micronuclei Test in Buccal Cells of Female Farmers and Children Exposed to Pesticides of Maneadero Agricultural Valley, Baja California, Mexico." *Journal of Toxicology*, 1-8. doi: 10.1155/2016/7934257.

Celik, A., Diler, S. B. & Eke, D. (2010). "Assessment of genetic damage in buccal epithelium cells of painters: micronucleus, nuclear changes, and repair index." *DNA Cell Biol*, *29* (6), 277-84. doi: 10.1089/dna.2009.0996.

Ceppi, M., Biasotti, B., Fenech, M. & Bonassi, S. (2010). "Human population studies with the exfoliated buccal micronucleus assay: statistical and epidemiological issues." *Mutat Res*, *705* (1), 11-9. doi: 10.1016/j.mrrev.2009.11.001.

Chang, W. P., Lee, S., Tu, J. & Hseu, S. (1996). "Increased micronucleus formation in nurses with occupational nitrous oxide exposure in operating theaters." *Environ Mol Mutagen, 27* (2), 93-7. doi: 10.1002/ (sici)1098-2280(1996)27:2<93::aid-em3>3.0.co;2-f.

da Silva Augusto, L. G., Lieber, S. R., Ruiz, M. A. & de Souza, C. A. (1997). "Micronucleus monitoring to assess human occupational exposure to organochlorides." *Environ Mol Mutagen, 29* (1), 46-52.

da Silva Junior, F., Tavella, R. A., Fernandes, C., Soares, M., de Almeida, K. A., Garcia, E. M., da Silva Pinto, E. A. & Baisch, A. (2018). "Genotoxicity in Brazilian coal miners and its associated factors." *Hum Exp Toxicol, 37* (9), 891-900. doi: 10.1177/0960327117745692.

Doherty, A. T., Ellard, S., Parry, E. M. & Parry, J. M. (1996). "A study of the aneugenic activity of trichlorfon detected by centromere-specific probes in human lymphoblastoid cell lines." *Mutat Res, 372* (2), 221-31.

Donbak, L., Rencuzogullari, E., Yavuz, A. & Topaktas, M. (2005). "The genotoxic risk of underground coal miners from Turkey." *Mutat Res, 588* (2), 82-7. doi: 10.1016/j.mrgentox.2005.08.014.

Falck, G. C., Hirvonen, A., Scarpato, R., Saarikoski, S. T., Migliore, L. & Norppa, H. (1999). "Micronuclei in blood lymphocytes and genetic polymorphism for GSTM1, GSTT1 and NAT2 in pesticide-exposed greenhouse workers." *Mutat Res, 441* (2), 225-37. doi: 10.1016/ s1383-5718(99)00051-0.

Fenech, M. (2006). "Cytokinesis-block micronucleus assay evolves into a "cytome" assay of chromosomal instability, mitotic dysfunction and cell death." *Mutat Res, 600* (1-2), 58-66. doi: 10.1016/ j.mrfmmm.2006.05.028.

Fenech, M., Bonassi, S., Turner, J., Lando, C., Ceppi, M., Chang, W. P., Holland, N., Kirsch-Volders, M., Zeiger, E., Bigatti, M. P., Bolognesi, C., Cao, J., De Luca, G., Di Giorgio, M., Ferguson, L. R., Fucic, A., Lima, O. G., Hadjidekova, V. V., Hrelia, P., Jaworska, A., Joksic, G., Krishnaja, A. P., Lee, T. K., Martelli, A., McKay, M. J., Migliore, L., Mirkova, E., Muller, W. U., Odagiri, Y., Orsiere, T., Scarfi, M. R., Silva, M. J., Sofuni, T., Surralles, J., Trenta, G., Vorobtsova, I., Vral, A. & Zijno, A. (2003). "Intra- and inter-laboratory variation in the scoring of micronuclei and nucleoplasmic bridges in binucleated human lymphocytes. Results of an international slide-scoring exercise by the HUMN project." *Mutat Res, 534* (1-2), 45-64.

Fenech, M., Holland, N., Chang, W. P., Zeiger, E. & Bonassi, S. (1999). "The HUman MicroNucleus Project--An international collaborative study on the use of the micronucleus technique for measuring DNA damage in humans." *Mutat Res, 428* (1-2), 271-83.

Fenech, M., Holland, N., Zeiger, E., Chang, W. P., Burgaz, S., Thomas, P., Bolognesi, C., Knasmueller, S., Kirsch-Volders, M. & Bonassi, S. (2011). "The HUMN and HUMNxL international collaboration projects on human micronucleus assays in lymphocytes and buccal cells--past, present and future." *Mutagenesis, 26* (1), 239-45. doi: 10.1093/mutage/geq051.

Fenech, M., Kirsch-Volders, M., Rossnerova, A., Sram, R., Romm, H., Bolognesi, C., Ramakumar, A., Soussaline, F., Schunck, C., Elhajouji, A., Anwar, W. & Bonassi, S. (2013). "HUMN project initiative and review of validation, quality control and prospects for further development of automated micronucleus assays using image cytometry systems." *Int J Hyg Environ Health, 216* (5), 541-52. doi: 10.1016/j.ijheh.2013.01.008.

Fenech, M. & Morley, A. (1985a). "Solutions to the kinetic problem in the micronucleus assay." *Cytobios, 43* (172-173), 233-46.

Fenech, M. & Morley, A. A. (1985b). "Measurement of micronuclei in lymphocytes." *Mutat Res, 147* (1-2), 29-36.

Heddle, J. A., Fenech, M., Hayashi, M. & MacGregor, J. T. (2011). "Reflections on the development of micronucleus assays." *Mutagenesis, 26* (1), 3-10. doi: 10.1093/mutage/geq085.

Holland, N., Bolognesi, C., Kirsch-Volders, M., Bonassi, S., Zeiger, E., Knasmueller, S. & Fenech, M. (2008). "The micronucleus assay in human buccal cells as a tool for biomonitoring DNA damage: the HUMN project perspective on current status and knowledge gaps." *Mutat Res, 659* (1-2), 93-108. doi: 10.1016/j.mrrev.2008.03.007. https://www.osha.gov/SLTC/lead/.

Huvinen, M., Mäkitie, A., Järventaus, H., Wolff, H., Stjernvall, T., Hovi, A., Hirvonen, A., Ranta, R., Nurminen, M. & Norppa, H. (2002). "Nasal cell micronuclei, cytology and clinical symptoms in stainless steel production workers exposed to chromium." *Mutagenesis, 17* (5), 425-429.

Joksic, G., Vidakovic, A. & Spasojevic-Tisma, V. (1997). "Cytogenetic monitoring of pesticide sprayers." *Environ Res, 75* (2), 113-8. doi: 10.1006/enrs.1997.3753.

Kirsch-Volders, M., Elhajouji, A., Cundari, E. & Hummelen, P. V. (1997). "The *in vitro* micronucleus test: a multi-endpoint assay to detect simultaneously mitotic delay, apoptosis, chromosome breakage, chromosome loss and non-disjunction." *Mutation Research/Genetic Toxicology and Environmental Mutagenesis, 392* (1), 19-30.

Leon-Mejia, G., Quintana, M., Debastiani, R., Dias, J., Espitia-Perez, L., Hartmann, A., Henriques, J. A. & Da Silva, J. (2014). "Genetic damage in coal miners evaluated by buccal micronucleus cytome assay." *Ecotoxicol Environ Saf, 107*, 133-9. doi: 10.1016/ j.ecoenv. 2014.05.023.

Lindholm, C., Norppa, H., Hayashi, M. & Sorsa, M. (1991). "Induction of micronuclei and anaphase aberrations by cytochalasin B in human lymphocyte cultures." *Mutat Res, 260* (4), 369-75.

Liu, Y., Sundqvist, K., Belinsky, S. A., Castonguay, A., Tjãlve, H. & Grafstrom, R. C. (1993). "Metabolism and macromolecular interaction of the tobacco-specific carcinogen 4-(methylnitrosamino)-1-(3-pyridyl)-1-butanone in cultured explants and epithelial cells of human buccal mucosa." *Carcinogenesis, 14* (11), 2383-2388.

Miller, R. C. (1973). "The micronucleus test as an *in vivo* cytogenetic method." *Environmental health perspectives, 6*, 167-170. doi: 10.1289/ehp.7306167.

Moon, J. D., Suh, S. P., Park, J. S., Cho, J. H. & Ahn, K. W. (1998). "Assessment of Genotoxic Hazard in Petrochemical Workers." *Korean Journal of Occupational and Environmental Medicine, 10* (1), 53-60.

Moore, L. E., Warner, M. L., Smith, A. H., Kalman, D. & Smith, M. T. (1996). "Use of the fluorescent micronucleus assay to detect the genotoxic effects of radiation and arsenic exposure in exfoliated human epithelial cells." *Environ Mol Mutagen*, *27* (3), 176-84. doi: 10.1002/(sici)1098-2280(1996)27:3<176::aid-em2>3.0.co;2-d.

Nersesyan, A., Kundi, M., Atefie, K., Schulte-Hermann, R. & Knasmuller, S. (2006). "Effect of staining procedures on the results of micronucleus assays with exfoliated oral mucosa cells." *Cancer Epidemiol Biomarkers Prev*, *15* (10), 1835-40. doi: 10.1158/1055-9965.epi-06-0248.

Norppa, H. & Falck, G. C. (2003). "What do human micronuclei contain?" *Mutagenesis*, *18* (3), 221-33. doi: 10.1093/mutage/18.3.221.

Organization for Economic Cooperation and Development (OECD). (2016). "*Mammalian Erythrocyte Micronucleus Test.*"

Palanikumar, L. & Panneerselvam, N. (2011). "Micronuclei assay: A potential biomonitoring protocol in occupational exposure studies." *Russian journal of genetics*, *47* (9), 1033.

Pastor, S., Creus, A., Parron, T., Cebulska-Wasilewska, A., Siffel, C., Piperakis, S. & Marcos, R. (2003). "Biomonitoring of four European populations occupationally exposed to pesticides: use of micronuclei as biomarkers." *Mutagenesis*, *18* (3), 249-58. doi: 10.1093/mutage/18.3.249.

Pinto, D., Ceballos, J. M., Garcia, G., Guzman, P., Del Razo, L. M., Vera, E., Gomez, H., Garcia, A. & Gonsebatt, M. E. (2000). "Increased cytogenetic damage in outdoor painters." *Mutat Res*, *467* (2), 105-11. doi: 10.1016/s1383-5718(00)00024-3.

RafiqKhan, M., Krishnan, G. T., Keezhekalam, R., Suresh, S. N., Pongiya, U. & Rao, Y. R. (2014). "Micronucleus assessment as a biomarker and susceptibility to DNA damage in workers occupationally exposed to pesticides." *Biomedical Research and Therapy*, *1*. doi: 10.7603/ s40730-014-0013-6.

Rohr, P., da Silva, J., da Silva, F. R., Sarmento, M., Porto, C., Debastiani, R., Dos Santos, C. E., Dias, J. F. & Kvitko, K. (2013). "Evaluation of genetic damage in open-cast coal mine workers using the buccal micronucleus cytome assay." *Environ Mol Mutagen*, *54* (1), 65-71. doi: 10.1002/em.21744.

Rosin, M. P. (1992). "The use of the micronucleus test on exfoliated cells to identify anti-clastogenic action in humans: a biological marker for the efficacy of chemopreventive agents." *Mutat Res*, *267* (2), 265-76.

Rosin, M. P. & German, J. (1985). "Evidence for chromosome instability *in vivo* in Bloom syndrome: increased numbers of micronuclei in exfoliated cells." *Hum Genet*, *71* (3), 187-91.

Rozgaj, R., Kasuba, V. & Jazbec, A. (2001). "Preliminary study of cytogenetic damage in personnel exposed to anesthetic gases." *Mutagenesis*, *16* (2), 139-43. doi: 10.1093/mutage/16.2.139.

Sarto, F., Finotto, S., Giacomelli, L., Mazzotti, D., Tomanin, R. & Levis, A. G. (1987). "The micronucleus assay in exfoliated cells of the human buccal mucosa." *Mutagenesis*, *2* (1), 11-7.

Spivack, Simon D., Gregory, J Hurteau., Ritu, Jain., Shalini, V Kumar., Kenneth, M. Aldous., John, F. Gierthy. & Laurence, S. Kaminsky. (2004). "Gene-environment interaction signatures by quantitative mRNA profiling in exfoliated buccal mucosal cells." *Cancer research*, *64* (18), 6805-6813.

Teixeira, J. P., Gaspar, J., Silva, S., Torres, J., Silva, S. N., Azevedo, M. C., Neves, P., Laffon, B., Mendez, J., Goncalves, C., Mayan, O., Farmer, P. B. & Rueff, J. (2004). "Occupational exposure to styrene: modulation of cytogenetic damage and levels of urinary metabolites of styrene by polymorphisms in genes CYP2E1, EPHX1, GSTM1, GSTT1 and GSTP1." *Toxicology*, *195* (2-3), 231-42.

Thomas, P., Harvey, S., Gruner, T. & Fenech, M. (2008). "The buccal cytome and micronucleus frequency is substantially altered in Down's syndrome and normal ageing compared to young healthy controls." *Mutat Res*, *638* (1-2), 37-47. doi: 10.1016/j.mrfmmm.2007.08.012.

Thomas, P., Holland, N., Bolognesi, C., Kirsch-Volders, M., Bonassi, S., Zeiger, E., Knasmueller, S. & Fenech, M. (2009). "Buccal micronucleus cytome assay." *Nat. Protocols*, *4* (6), 825-837. doi: http://www.nature.com/nprot/journal/v4/n6/suppinfo/nprot.2009.53_S1 .html.

Tolbert, P. E., Shy, C. M. & Allen, J. W. (1992). "Micronuclei and other nuclear anomalies in buccal smears: methods development." *Mutat Res*, *271* (1), 69-77. doi: 10.1016/0165-1161(92)90033-i.

Ulker, O. C., Ustundag, A., Duydu, Y., Yucesoy, B. & Karakaya, A. (2008). "Cytogenetic monitoring of coal workers and patients with coal workers' pneumoconiosis in Turkey." *Environ Mol Mutagen*, *49* (3), 232-7. doi: 10.1002/em.20377.

Vaglenov, A., Carbonell, E. & Marcos, R. (1998). "Biomonitoring of workers exposed to lead. Genotoxic effects, its modulation by polyvitamin treatment and evaluation of the induced radioresistance." *Mutat Res*, *418* (2-3), 79-92. doi: 10.1016/s1383-5718(98)00111-9.

Vaglenov, A., Creus, A., Laltchev, S., Petkova, V., Pavlova, S. & Marcos, R. (2001). "Occupational exposure to lead and induction of genetic damage." *Environ Health Perspect*, *109* (3), 295-8. doi: 10.1289/ehp.01109295.

Vondracek, M., Xi, Z., Larsson, P., Baker, V., Mace, K., Pfeifer, A., Tjälve, H., Donato, M. T., Gomez-Lechon, M. J. & Grafström, R. C. (2001). "Cytochrome P450 expression and related metabolism in human buccal mucosa." *Carcinogenesis*, *22* (3), 481-488.

Chapter 5

AN OVERVIEW ON MICRONUCLEUS ASSAY EVALUATION BY FLOW CYTOMETRY

María Sánchez-Flores[1,2,3], PhD, Eduardo Pásaro[1], PhD, Blanca Laffon[1,*], PhD and Vanessa Valdiglesias[1,3], PhD

[1]Department of Psychology, Area of Psychobiology, DICOMOSA Group, Universidade da Coruña, A Coruña, Spain
[2]Environmental Health Department, National Health Institute, Porto, Portugal
[3]EPIUnit – Instituto de Saúde Pública, Universidade do Porto, Porto, Portugal

ABSTRACT

Micronucleus (MN) assay is a commonly used method to evaluate chromosome alterations. MN are expressed in dividing cells as the result of chromosome fragments or whole chromosomes that lag behind during anaphase. Therefore, MN assay provide a reliable measure of both chromosome breakage and chromosome loss, and thus MN frequency is a widely accepted biomarker of genotoxicity and genomic instability. Traditionally, the method most commonly employed for MN assessment

---

[*] Corresponding Author's Email: blaffon@udc.es.

is microscopy. However, this technique is time consuming, highly subjective, and the number of cells scored is relatively low. In this context, flow cytometry, as a high throughput alternative method, allows the automation of the MN scoring to overcome the mentioned limitations of the standard microscopy scoring. Besides, in the last years, several imaging flow cytometry platforms have been developed to evaluate the MN assay, which enable the capture of high resolution images in addition to the traditional flow cytometry features. This chapter presents an overview of the evaluation of MN by flow cytometry, reviewing the studies published in the international literature so far that employ different experimental designs (*in vitro*, *in vivo*, human biomonitoring) for a variety of purposes (radiation biodosimetry, risk assessment, aneugen and clastogen classification, etc.).

INTRODUCTION

Microcnuclei (MN) are small, extranuclear chromatin bodies enclosed in a nuclear envelop. They originate during cell division from acentric chromosome fragments or whole chromosomes unable to migrate to the mitotic spindle poles (Avlasevich et al., 2011). Therefore, the presence of these bodies in the cell cytoplasm is indicative of chromosome alterations (Fenech et al., 2007).

The mechanisms by which MN are induced are well known. Clastogenic agents induce chromosome breakage, while aneugenic agents cause the missegregatation of whole chromatids or chromosomes during cell division, both of them leading to MN formation. Hence, the MN assay allows to asses both chromosome breakage and chromosome loss in a reliable way. In this frame, MN are commonly used as an endpoint for *in vivo* and *in vitro* genotoxicity testing (Hintzsche et al., 2017), what led to the development of OECD (Organization for Economic Cooperation and Development) guidelines for an appropriate use of this assay for both purposes (OECD Test No. 474 and Test No. 487, 2016, respectively).

MN are also frequently used in human biomonitoring as an indicator of *in vivo* occupational or environmental exposure to genotoxic agents, as well as a biomarker of genomic instability in epidemiological studies (Rodrigues et al., 2018). Indeed, MN frequency in human lymphocytes has been proposed as a predictor of cancer risk (Bonassi et al., 2007; Pardini et al., 2017; Podrimaj-Bytyqi et al., 2018).

Traditionally, MN have been evaluated by visual observation of the cells in the microscope. The cytokinesis-block micronucleus (CBMN) assay, firstly described by Fenech and Morley in 1985, allows to identify cells that have experienced one cell division due to their appearance as binucleated cells after blocking cytokinesis with cytochalasin-B (Cyt-B), an inhibitor of microfilament ring assembly required for the completion of cytokinesis (Fenech, 2007). Visual inspection also enables the detection of other alterations, such as nuclear buds and nucleoplasmic bridges, as well as polynucleated cells (Fenech, 2007). However, visual microscope-based scoring is tedious, time consuming and prone to scorer subjectivity, leading to inter- and intra-scorer variability, due to fatigue when many slides need to be scored (Rodrigues, 2018). Besides, the number of cells evaluated in slide preparations is relatively small (Witt et al., 2008).

In the last decades, in order to overcome the aforementioned limitations of visual scoring, there have been several attempts towards the automation of the MN scoring. Automated microscope scoring methods [Metafer™ MNScore (Metasystems, GmbH Altlussheim, Germany), PathFinder™ Cellscan™ (IMSTAR, Paris, France), and Compucyte iCyte® laser scanning cytometer (Thorlabs, Sterling VA, USA)], with software algorithms capable of identifying and capturing images of fluorescently-labeled nuclei and MN, have increased sample throughput when compared to manual scoring methods (Decordier et al., 2011; François et al., 2014; Rossnerova et al., 2011). These methods also enable the storage of images for data re-evaluation.

However, these methods still have several limitations like the requirement of lengthy and/or complex sample processing protocols or the need to elaborate high quality microscope slides. Besides, the lack of staining consistency and suboptimal contrast between nucleus and cytoplasm can lead to high rates of false positives (reviewed in Rodrigues et al., 2018). For this reason, a visual verification of the image gallery is necessary, which reduces the method throughput and can result in scorer bias (Rodrigues, 2018).

A first automatic measurement of MN by flow cytometry (FCM) was proposed by Nüsse and Kramer (1984). These authors developed a two step method to separate MN from the main nucleus: (i) treatment of the cells with a detergent solution to lyse the cell membrane; (ii) removal of the cytoplasm with a citric acid-sucrose solution. Subsequently, MN were identified based on ethidium bromide fluorescence and quantifying differences in DNA intensity.

However, this technique presented a major drawback: it was not able to distinguish MN from the chromatin of necrotic and mid/late apoptotic bodies, leading to false positives. Avlasevich et al. (2006) proposed the use of ethidium monoazide (EMA), a fluorescent photoaffinity label that, after photolysis, binds covalently to nucleic acids of dead and dying cells. Subsequently, cells are stained with the nucleic acid dye SYTOX green. This protocol results in differential staining of chromatin from healthy cells (EMA-/SYTOX+) relative to necrotic and late stage apoptotic cells, which show a EMA+/SYTOX+ profile. This sequential staining approach allows identifying and excluding the necrotic and apoptotic subcellular particles that could be mistaken as MN.

The use of FCM in the evaluation of the MN assay presents several advantages: (i) the ability to rapidly analyze large numbers of cells in short times (thousands of cells in few minutes), which reduces sampling variability and enhances statistical power; (ii) the objectivity of the method; (iii) the demonstrated correlation with the conventional microscopy evaluation; (iv) furthermore, the possibility of sorting the MN by FCM paired with the FISH-technique provides a tool for measuring chromosomal content (Avlasevich et al., 2006; Bemis et al., 2016; Nüsse and Marx, 1997; Rodrigues et al., 2018; Witt et al., 2008).

The staining and analyzing techniques for FCM have been improved in the last years (Bryce et al., 2007), to the point that Litron Laboratories (Rochester, NY, USA) has developed commercially available kits for the evaluation of the MN assay (Collins et al., 2008).

Despite of all the advantages of the MN assay analysis by FCM, the technique still presents some limitations. Because of the requirement of cell lysing, Cyt-B is useless to be employed, therefore binucleated and polinucleated cells cannot be assessed and it is not possible to know if cells have undergone one cell division. Besides, lysing the cells makes it impossible to relate MN to individual cells; only the total number of MN in the cell suspension can be determined, but not the number of micronucleated cells or the number of MN in each cell. This can underestimate the level of chromosome damage (Smolewski et al., 2001). Besides, the cell lysis originates debris that can interfere in the analysis. In addition, MN legitimacy cannot be visually confirmed and, unlike in automated microscopy, data cannot be re-evaluated (Rodrigues, 2018).

To overcome these limitations, imaging flow cytometry was developed as a system combining the high resolution imagery obtained by microscopy scoring with the high-throughput nature of the conventional FCM (Rodrigues et al., 2014). This method, similarly to the traditional FCM, consists in labelling cells with fluorescent dyes. The imaging flow cytometer simultaneously captures fluorescent and brightfield images, enabling different cellular structures to be analyzed (Wang et al., 2019).

In addition to the typical advantages of the automated microscopy (storage of all the imagery for possible re-analyses) and FCM scoring (high throughput data acquisition), imaging flow cytometry not only allows to visualize the cell cytoplasm but also other cellular structures such as nuclei, centromeres or telomeres. However, this scoring methodology presents as well some limitations: (i) the MN frequencies obtained are lower than those from visual microscopy (i.e., 30% in radiation dosimetry); (ii) two dimensional images of three dimensional cells can limit the visibility of some MN; (iii) small MN or MN close to the main nucleus can be missed by the mathematical algorithms used to perform image analysis (Rodrigues, 2018; Wang et al., 2019).

The next sections present an overview on the use of MN assay evaluation by FCM in *in vitro* and *in vivo* studies in different cell lines and organisms, as well as in human biomonitoring studies, for several purposes (radiation biodosimetry, risk assessment, aneugen and clastogen classification, etc.).

IN VITRO STUDIES

As previously mentioned, the *in vitro* FCM MN assay is commonly used to evaluate genotoxicity in mammalian cells, primarily using suspension cell cultures and human lymphocytes but also with attached cell lines (e.g., CHO-K1 and V-79 cells) routinely used in the *in vitro* microscopic evaluation of MN (Bryce et al., 2010a; Kligerman et al., 2015; Lukamowicz et al., 2011; Nüsse and Marx, 1997).

Several studies suggest that, for certain cell lines, FCM data may provide information about genotoxic mechanism of action signatures that are able to discriminate between clastogenic and aneugenic activities. Bryce et al. (2011) were able to classify the 16 chemicals studied as clastogenic or aneugenic according to the %MN, %hyplodiploid nuclei and the median SYTOX fluorescence intensity of MN events in CHO-K1 cells.

In this frame, pharmaceutic industries frequently use the *in vitro* FCM MN assay in cell lines such as TK6 or V79 cells as part of the regulatory test battery to assess drug candidates for their ability to induce genotoxicity (Lukamowicz et al., 2011; Nicolette et al., 2010). Consequently, positive compounds require follow-up testing strategies to determine their mode of action, such as the use of the aneugenicity markers phosphorylated-histone 3 (p-H3) and polyploidy, the clastogenicity marker phosphorylated histone 2AX (H2AX), and the apoptosis marker cleaved caspase 3, among others (Bryce et al., 2014; Cheung et al., 2015).

The FCM methodology for scoring MN *in vitro* is also described as a practical and efficient approach for characterizing genotoxicity dose-response relationships in human TK-6 cells (Bryce et al., 2010b; Buick et al., 2017).

The FCM MN frequency is used in studies for risk assessment. Allemang et al. (2018) studied the capability of fifteen pyrrolizidine alkaloids, which can be found in food from plant origin, to induce DNA damage in HepaRG human liver cells. Ali et al. (2011) evaluated the MN induction by the fungal toxin ochratoxin A in comparison with the ability of this toxin to induce cytotoxicity and DNA damage in two mammalian cell lines, CHO-K1-BH4 and TK6 cells. This toxin is a common contaminant of human food and animal feed that can cause nephropathy in animals and urinary tract tumors in humans. Abramsson-Zetterberg et al. (2010) found no genotoxic effect analyzig the frequency of MN in human lymphocytes treated *in vitro* with the toxin microcystin-LR and cyanobacterial extracts. Okadaic acid is the main representative of diarrheic shellfish poisoning toxin, and it is reported to induce MN, evaluated by the traditional microscopic technique, in CHO cells (Le Hégarat et al., 2006). Valdiglesias et al. (2011) found increased MN rates in SHSY5Y and HepG2 cells treated with okadaic acid, but not in human peripheral leukocytes.

The use of the *in vitro* estimation of FCM MN frequency in RTG-2 cells (rainbow trout-derived cells) was suggested as a possible screening test for the genotoxicity of industrial waste waters (Kohlpoth et al., 1999) and sewage treatment plant effluents (Llorente et al., 2012), and as an alternative for environmental risk assessment in fish populations (Castaño et al., 2000).

Use of nanomaterials in consumer products has increased greatly in recent years. For this reason, screening tools for nanoparticle (NP) safety evaluation are necessary. For nanotoxicology, the FCM approach offers the possibility to analyze higher doses, limited in the conventional microscopic scoring due to the difficulties in distinguishing fluorescent particle agglomerates from MN (Di Bucchianico et al., 2017).

Silver NP are one of the most often used NP in food and cosmetic products, due to their antibacterial and antifungal properties. Butler et al. (2015) studied the clastogenicity of silver NP in the human cell lines Jurkat (clone E6-1) and THP-1. Sahu et al. (2014; 2016) evaluated the genotoxic potential of food-related nanosilver exposure in hepatoblastoma HepG2 and colon carcinoma Caco2 cells. Jiang et al. (2013) investigated and determined the suitability of CHO-K1 cells for genotoxicity testing of silver nanoparticles.

Genotoxicity of titanium dioxide NP has been also assessed employing the *in vitro* MN assay by FCM in several cell lines like BEAS-2B (Di Bucchianico et al., 2017), HepG2 (Vallabani et al., 2014), SHSY5Y (Valdiglesias et al., 2013a), and A549 (Jugan et al., 2012), showing a significant correlation with the traditional microscopy methodology in some cases and discrepancies in others.

Zinc oxide nanoparticles (ZnO NP) are widely used in different areas such as cosmetics, medicine, coatings, paints, electronics, catalysis and as an antibacterial agent. Consequently, assessing the mutagenic potential of these nanoparticles is of paramount importance. Jain et al. (2019) evaluated the induction of MN in Chinese hamster lung fibroblast cells (V-79) as an *in vitro* model exposed to ZnO NP. A significant increase in the FCM MN frequency was observed in a concentration-dependent manner. Besides, Valdiglesias et al. (2013b) observed no induction of MN in human neuronal cells (SH-SY5Y) treated with ZnO NP for 3 h, but an important dose-dependent MN induction was observed after a 6 h treatment.

Due to their particular superparamagnetic properties, iron oxide nanoparticles have a variety of biomedical applications. No MN induction by oleic acid-coated or silica-coated iron oxide nanoparticles was observed in SH-SY5Y cells (Fernández-Bertólez et al., 2018a; Kilic et al., 2015). Similarly, negative results were obtained for silica-coated iron oxide particles in glial cells (Fernández-Bertólez et al., 2018b).

IN VIVO STUDIES

The use of the mouse MN test is widespread as a short-term *in vivo* system to screen chemicals for genotoxic activity. There were several early attempts at automatically scoring the frequency of micronucleated erythrocytes by FCM (Hayashi et al., 1992; Hutter and Stöhr, 1982; Tometsko et al., 1993a; 1993b; 1993c). However, the methodologies proposed were not able to separate analysis of reticulocytes (RET), responsive to acute genotoxicant exposures, and mature red blood cells. In 1992, Grawé et al. proposed a dual-dye (Hoechst 33342/thiazole orange) that enabled the measure of the frequency of micronucleated reticulocytes (MN-RET) in mouse blood by FCM. Later on, a simpler and reliable method based on single-laser flow cytometry using CD-71 (transferrin receptor) antibodies was developed and optimized by Dertinger et al. (1996, 2000, 2002, 2004). Afterwards, Balmus et al. (2015) scaled down Dertinger's FCM method to perform the assay in 96-well plates for the assessment of genome stability in mice.

In the following years, modifications of Dertinger's methodology emerged to enable its use in other species of toxicological interest, such as zebra fish (Le Bihanic et al., 2016), rats (MacGregor et al., 2006), beagle dogs (Harper et al., 2007; McKeon et al., 2012) and non-human primates (rhesus monkey) (Hotchkiss et al., 2008; Morris et al., 2009), as well as in human blood (Abramsson-Zetterberg et al., 2000).

The *in vivo* evaluation of MN-RET in peripheral blood by FCM has been used in exposure studies to assess the protective effect against DNA damage (Gradecka-Meesters et al., 2011) or the genotoxicity in rodents of compounds that can be found: (i) in food, such as glycidol (Aasa et al., 2017; Dobrovolsky et al., 2016), acrylamide (Abramsson-Zetterberg et al., 2005; Dobrovolsky et al., 2016; Ghanayema et al., 2005; Paulsson et al., 2002; Zeiger et al., 2009), furan (Durling et al., 2007), phytosterols (Abramsson-Zetterberg et al., 2007), cyanobacterial extracts and microcystin-LR (Abramsson-Zetterberg et al., 2010), semicarbazide (Abramsson-Zetterberg and Svensson, 2005), 4-methylbenzophenone and benzophenone (Abramsson-Zetterberg and Svensson, 2011), food coloring agent Allura Red AC (E129) (Abramsson-Zetterberg and Ilbäck, 2013); (ii) in cosmetics and perfumes, such as macrocyclic musk compounds (Abramsson-Zetterberg and Slanina, 2002); (iii) in drugs used in chemotherapy, such as doxorubicin (Manjanatha et al., 2014), or in the treatment of infections, such as the combination of trimethoprim and sulfamethoxazole (Ortiz et al., 2011); (iv) in plastics, such as styrene-acrylonitrile trimer (Hobbs et al., 2012); and (v) in industrial chemicals, such as 1,3-butadiene (Lahdetie et al., 1997) or formadehyde (Speit et al., 2009).

The FCM-based MN assay has also been used to identify possible thresholds in dose-response genotoxicity curves, e.g., for aneuploidy caused by spindle poisons in mice (Cammerer et al., 2009) and for clastogenic and aneugenic compounds in rats (Asano et al., 2006).

Several authors have tested the genotoxicity of different agents by means of the FCM MN-RET frequency paired with the Pig-a assay, which measures gene mutation in the red blood cells, in mice (Bhalli et al., 2011;

Labash et al., 2016; Phonethepswath et al., 2013) and most commonly in rats (Avlasevich et al., 2014; Cammerer et al., 2007, 2011; Dertinger et al., 2011, 2014; Dobrovolsky et al., 2010; Elhajouji et al., 2018; Guérard et al., 2013; Shi et al., 2011; Stankowski et al., 2011; Tu et al., 2015; Zhou et al., 2014).

The *in vivo* FCM MN assay was also employed to evaluate the effect of exposure to electromagnetic fields on the reproductive system of male rats (Kumar et al., 2012). Besides, FCM scoring of micronucleated reticulocytes has been proposed as a possible high-throughput radiation biodosimeter and has been used to evaluate the genotoxic effects of radiation on mice. Liu et al. (2009) observed a linear dose-response relationship in peripheral blood MN-RET in mice exposed to X-ray. Graupner et al. (2017) used the levels of MN-RET to compare the genotoxic effects of low dose rate (gamma radiation) and high dose rate (X-rays) radiations in mice to study the development of colon cancer after chronic exposure to these radiations. Hamasaki et al. (2007) observed delayed genotoxic effects of irradiation by means of significantly increased frequencies of MN-RET in mice even a year after the exposure to X-rays.

HUMAN STUDIES

The frequency of MN-RET in human blood by FCM is an endpoint used for: (i) assessing genotoxic potential of pharmaceuticals undergoing clinical trials; (ii) identifying polymorphisms or epigenetic effects that may lead to hypersensitivity to particular genotoxicants; (iii) evaluating compounds, diet elements or other factors that may protect against endogenous- or xenobiotic-induced genetic damage; (iv) determining the level of DNA damage in a population following accidental exposures, and (v) biomonitoring workers occupationally exposed to potentially genotoxic materials (Dertinger et al., 2010).

Harrod et al. (2007) used FCM to study the levels of Howell-Jolly bodies or MN in blood reticulocytes from children with sickle cell anemia, while Flanagan et al. (2010) employed the frequency of MN-RET by FCM to study the genotoxic safety of hydroxyurea in the treatment of the same disease in children. Even though hydroxyurea is considered as non-DNA reactive, the authors observed that clinically relevant exposure to this compound resulted in an increase of MN-RET frequency. However, the results also showed an important inter-patient variability. In a similar way, Grawé et al. (2005) evaluated the level of micronucleated transferrin-receptor positive reticulocytes (MN-Trf-Ret) by FCM in young patients treated with ^{131}I for thyroid cancer. The observed MN-Tf-Ret frequency increased within 1 day to a maximum and declined in the following 2–5 days to its value before treatment. These results supported the use of this method for monitoring individuals after a radiation accident.

Pavlyushchik et al. (2016) evaluated the association between the angiotensin-converting enzyme ACE I/D polymorphism (rs 4340) and DNA damage in patients with essential hypertension in men. They observed an increase in MN in only the carriers of hypertensive II genotype as compared to normotensive individuals.

Abramsson-Zetterberg et al. (2006) studied the possible correlation between the MN level, in human MN-Trf-Ret, and folate status in non-folate-deficient subjects. With this aim, they conducted three cross-sectional studies (N = 32, 29, and 38, respectively), two of them connected to dietary interventions, to clarify whether nutritional supplementations (folic acid) had any effect on the MN-Trf-Ret frequency, evaluated by FCM, as a biomarker of chromosome stability. Their results showed a positive correlation between high folate status and high chromosome stability. Kotova et al. (2015) employed FCM to study the possible influence of a vegetarian/non-vegetarian diet on genotoxic effects. Their findings, lower frequency of MN-Trf-Ret in vegetarians, might suggest a beneficial effect of this diet on genomic stability in healthy individuals.

The relationship between malnutrition and genomic instability, and its association with a bacterial infection in children, was studied by Cervantes-Ríos et al. (2012). The frequency of MN-RET evaluated by FCM showed an increase associated with bacterial infection and with malnutrition.

Laffon et al. (2014) evaluated the frequency of MN in lymphocytes by FCM in a follow-up study on the genotoxic effects in subjects exposed to an oil spill seven years after the exposure, to determine the possible persistence of genotoxic damage. They found a significantly lower MN frequency in the exposed subjects with regard to the control population, similar to the mononucleated micronucleated cells evaluated in the CBMN test by microscopy.

García-Lestón et al. (2012) used the FCM MN assay in leukocytes as a biomarker of genotoxicity in workers exposed to lead. Their results showed increased levels of of MN frequency in workers with regard to the unexposed controls.

CONCLUSION

This chapter summarize the basis of the MN assay and the several attempts to automate the scoring process, mainly to reduce the subjectivity of the traditional visual microscopy scoring and to obtain a high throughput technology that enables the scoring of larger number of cells and samples in a simpler and faster way. Although automated system for the analyses of MN have been developed, they still present some limitations.

While automated microscopy scoring allows to increase the throughput, in most cases visual verification of the image gallery is necessary, which may reduce throughput and may re-introduce scorer bias. Flow cytometry seems to solve this limitation, since there is no influence of the scorer, at the same time that thousands of cells can be scored in a few minutes. However, the impossibility of using the CBMN assay reduces the endpoints that can measured with this technique. On the other hand, image flow cytometry seems to include the advantages of the other two automated methods (objectivity and the possibility to perform the CBMN assay). Nevertheless, MN frequencies obtained by image flow cytometry show discrepancies with the ones obtained with visual microscopy scoring.

The FCM MN assay was suggested to be especially useful in the *in vitro* assessment of nanomaterial genotoxicity. Since some NP can interfere with image-based automated methodologies due to their fluorescence and their tendency to form agglomerates that can be mistaken with MN, the doses of NP that can be tested for MN induction are restricted to low concentrations.

However, FCM enables testing higher doses without methodological interference.

The development of an *in vivo* FCM MN assay in peripheral blood reticulocytes in mice, with similar results to those observed in bone marrow, was an important step for applying the scoring technique to other model animals. The use of *in vivo* blood samples instead of bone marrow eliminates the need to kill the animal, what means a reduction in the use of animals in preclinical safety studies.

Besides, due to the large number of samples that are usually analyzed in human biomonitoring studies, the FCM MN assay provides a more suitable method for these approaches, since it is faster and less complex than the microscopy-based methodology.

In this chapter, an overview of the use of the FCM-based MN assay was conducted. In brief, the FCM MN assay has been proven to be useful in genotoxicity testing of potential pharmaceutical compounds, new chemicals released to the market, food components, risk assessment, occupational and environmental exposure, and human biomonitoring.

ACKNOWLEDGMENTS

This work was supported by Xunta de Galicia (grant number ED431B 2019/02). V. Valdiglesias was supported by a Xunta de Galicia postdoctoral fellowship (reference ED481B 2016/190-0).

REFERENCES

Aasa, J., Abramsson-Zetterberg, L., Carlsson, H. & Törnqvist, M. (2017). The genotoxic potency of glycidol established from micronucleus frequency and hemoglobin adduct levels in mice. *Food and Chemical Toxicology.*, *100*, 168-174.

Abramsson-Zetterberg, L., Durling, L., Yang-Wallentin, F., et al. (2006). The impact of folate status and folic acid supplementation on the micronucleus frequency in human erythrocytes. *Mutation Research.*, *603*, 33-40.

Abramsson-Zetterberg, L. & Ilbäck, N. G. (2013). The synthetic food colouring agent Allura Red AC (E129) is not genotoxic in a flow cytometry-based micronucleus assay *in vivo*. *Food and Chemical Toxicology.*, *59*, 86-89.

Abramsson-Zetterberg, L. & Slanina, P. (2002). Macrocyclic musk compounds--an absence of genotoxicity in the Ames test and the *in vivo* Micronucleus assay. *Toxicology Letters.*, *135*, 155-163.

Abramsson-Zetterberg, L., Sundh, U. B. & Mattsson, R. (2010). Cyanobacterial extracts and microcystin-LR are inactive in the micronucleus assay *in vivo* and *in vitro*. *Mutation Research.*, *699*, 5-10.

Abramsson-Zetterberg, L. & Svensson, K. (2005). Semicarbazide is not genotoxic in the flow cytometry-based micronucleus assay *in vivo*. *Toxicology Letters.*, *155*, 211-217.

Abramsson-Zetterberg, L. & Svensson, K. (2011). 4-Methylbenzophenone and benzophenone are inactive in the micronucleus assay. *Toxicology Letters.*, *201*, 235-239.

Abramsson-Zetterberg, L., Svensson, M. & Johnsson, L. (2007). No evidence of genotoxic effect *in vivo* of the phytosterol oxidation products triols and epoxides. *Toxicology Letters.*, *173*, 132-139.

Abramsson-Zetterberg, L., Wong, J. & Ilbäck, N. G. (2005). Acrylamide tissue distribution and genotoxic effects in a common viral infection in mice. *Toxicology.*, *211*, 70-76.

Abramsson-Zetterberg, L., Zetterberg, G., Bergqvist, M. & Grawé, J. (2000). Human cytogenetic biomonitoring using flow-cytometric analysis of micronuclei in transferrin-positive immature peripheral blood reticulocytes. *Environmental and Molecular Mutagenesis.*, *36*, 22–31.

Ali, R., Mittelstaedt, R. A., Shaddock, J. G., et al. (2011). Comparative analysis of micronuclei and DNA damage induced by Ochratoxin A in two mammalian cell lines. *Mutation Research.*, *723*, 58-64.

Allemang, A., Mahony, C., Lester, C. & Pfuhler, S. (2018). Relative potency of fifteen pyrrolizidine alkaloids to induce DNA damage as measured by micronucleus induction in HepaRG human liver cells. *Food and Chemical Toxicology.*, *121*, 72-81.

Asano, N., Torous, D. K., Tometsko, C. R., et al. (2006). Practical threshold for micronucleated reticulocyte induction observed for low doses of mitomycin C, Ara-C and colchicine. *Mutagenesis.*, *21*, 15-20.

Avlasevich, S., Bryce, S., De Boeck, M., et al. (2011). Flow cytometric analysis of micronuclei in mammalian cell cultures: past, present and future. *Mutagenesis.*, *26*, 147-152.

Avlasevich, S. L., Bryce, S. M., Cairns, S. E. & Dertinger, S. D. (2006). *In vitro* micronucleus scoring by flow cytometry: differential staining of micronuclei versus apoptotic and necrotic chromatin enhances assay reliability. *Environmental and Molecular Mutagenesis.*, *47*, 56-66.

Avlasevich, S. L., Phonethepswath, S., Labash, C., et al. (2014). Diethylnitrosamine genotoxicity evaluated in sprague dawley rats using pig-a mutation and reticulocyte micronucleus assays. *Environmental and Molecular Mutagenesis.*, *55*, 400-406.

Balmus, G., Karp, N. A., Ng, B. L., et al. (2015). A high-throughput *in vivo* micronucleus assay for genome instability screening in mice. *Nature Protocols.*, *10*, 205-215.

Bemis, J. C., Wills, J. W., Bryce, S. M., et al. (2016). Comparison of *in vitro* and *in vivo* clastogenic potency based on benchmark dose analysis of flow cytometric micronucleus data. *Mutagenesis.*, *31*, 277-285.

Bhalli, J. A., Pearce, M. G., Dobrovolsky, V. N. & Heflich, R. H. (2011). Manifestation and persistence of Pig-a mutant red blood cells in C57BL/6 mice following single and split doses of N-ethyl-N-nitrosourea. *Environmental and Molecular Mutagenesis.*, *52*, 766-773.

Bonassi, S., Znaor, A., Ceppi, M., et al. (2007). An increased micronucleus frequency in peripheral blood lymphocytes predicts the risk of cancer in humans. *Carcinogenesis.*, *28*, 625–631.

Bryce, S. M., Avlasevich, S. L., Bemis, J. C. & Dertinger, S. D. (2011). Miniaturized flow cytometry-based cho-k1 micronucleus assay discriminates aneugenic and clastogenic modes of action. *Environmental and Molecular Mutagenesis.*, *52*, 280-286.

Bryce, S. M., Avlasevich, S. L., Bemis, J. C., et al. (2010). Miniaturized flow cytometric *in vitro* micronucleus assay represents an efficient tool for comprehensively characterizing genotoxicity dose-response relationships. *Mutation Research.*, *703*, 191-199.

Bryce, S. M., Bemis, J. C., Avlasevich, S. L. & Dertinger, S. D. (2007). *In vitro* micronucleus assay scored by flow cytometry provides a comprehensive evaluation of cytogenetic damage and cytotoxicity. *Mutation Research - Genetic Toxicology and Environmental Mutagenesis.*, *630*, 78–91.

Bryce, S. M., Bemis, J. C., Mereness, J. A., et al. (2014). Interpreting *in vitro* micronucleus positive results: simple biomarker matrix discriminates clastogens, aneugens, and misleading positive agents. *Environmental and Molecular Mutagenesis.*, *55*, 542-555.

Bryce, S. M., Shi, J., Nicolette, J., et al. (2010a). High content flow cytometric micronucleus scoring method is applicable to attachment cell lines. *Environmental and Molecular Mutagenesis.*, *51*, 260-266.

Buick, J. K., Williams, A., Kuo, B., et al. (2017). Integration of the TGx-28.65 genomic biomarker with the flow cytometry micronucleus test to assess the genotoxicity of disperse orange and 1,2,4-benzenetriol in human TK6 cells. *Mutation Research.*, *806*, 51-62.

Butler, K. S., Peeler, D. J., Casey, B. J., et al. (2015). Silver nanoparticles: correlating nanoparticle size and cellular uptake with genotoxicity. *Mutagenesis.*, *30*, 577-591.

Cammerer, Z., Bhalli, J. A., Cao, X., et al. (2011). Report on stage III Pig-a mutation assays using N-ethyl-N-nitrosourea-comparison with other *in vivo* genotoxicity endpoints. *Environmental and Molecular Mutagenesis.*, *52*, 721-730.

Cammerer, Z., Elhajouji, A. & Suter, W. (2007). *In vivo* micronucleus test with flow cytometry after acute and chronic exposures of rats to chemicals. *Mutation Research.*, *626*, 26-33.

Cammerer, Z., Schumacher, M. M., Kirsch-Volders, M., et al. (2010). Flow cytometry peripheral blood micronucleus test *in vivo*: determination of potential thresholds for aneuploidy induced by spindle poisons. *Environmental and Molecular Mutagenesis.*, *51*, 278-284.

Castaño, A., Sanchez, P., Llorente, M. T., et al. (2000). The use of alternative systems for the ecotoxicological screening of complex mixtures on fish populations. *The Science of Total Environment.*, *247*, 337-348.

Cervantes-Ríos, E., Ortiz-Muñiz, R., Martínez-Hernández, A. L., et al. (2012). Malnutrition and infection influence the peripheral blood reticulocyte micronuclei frequency in children. *Mutation Research.*, *731*, 68-74.

Cheung, J. R., Dickinson, D. A., Moss, J., et al. (2015). Histone markers identify the mode of action for compounds positive in the TK6 micronucleus assay. *Mutation Research Genetic Toxicology and Environmental Mutagenesis.*, *777*, 7-16.

Collins, J. E., Ellis, P. C., White, A. T., et al. (2008). Evaluation of the Litron *In Vitro* MicroFlow® Kit for the flow cytometric enumeration of micronuclei (MN) in mammalian cells. *Mutation Research.*, *654*, 76-81.

Decordier, I., Papine, A., Vande Loock, K., et al. (2011). Automated image analysis of micronuclei by IMSTAR for biomonitoring. *Mutagenesis.*, *26*, 163-168.

Dertinger, S. D., Camphausen, K., MacGregor, J. T., Bishop, M. E., Torous, D. K., Avlasevich, S., et al. (2004). Three-color labeling method for flow cytometric measurement of cytogenetic damage in rodent and human blood. *Environmental and Molecular Mutagenesis.*, *44*, 427–435.

Dertinger, S. D., Phonethepswath, S., Avlasevich, S. L., et al. (2014). Pig-a gene mutation and micronucleated reticulocyte induction in rats exposed to tumorigenic doses of the leukemogenic agents chlorambucil, thiotepa, melphalan, and 1,3-propane sultone. *Environmental and Molecular Mutagenesis.*, *55*, 299-308.

Dertinger, S. D., Phonethepswath, S., Weller, P., et al. (2011). Interlaboratory pig-a gene mutation assay trial: studies of 1,3-propane sultone with immunomagnetic enrichment of mutant erythrocytes. *Environmental and Molecular Mutagenesis.*, *52*, 748-755.

Dertinger, S. D., Torous, D. K., Hall, N. E., et al. (2000). Malaria-infected erythrocytes serve as biological standards to ensure reliable and consistent scoring of micronucleated erythrocytes by flow cytometry. *Mutation Research.*, *464*, 195-200.

Dertinger, S. D., Torous, D. K., Hall, N. E., et al. (2002). Enumeration of micronucleated CD71-positive human reticulocytes with a single-laser flow cytometer. *Mutation Research.*, *515*, 3-14.

Dertinger, S. D., Torous, D. K., Hayashi, M. & MacGregor, J. T. (2010). Flow cytometric scoring of micronucleated erythrocytes: an efficient platform for assessing *in vivo* cytogenetic damage. *Mutagenesis.*, *26*, 139-145.

Dertinger, S. D., Torous, D. K. & Tometsko, K. (1996). Simple and reliable enumeration of micronucleated reticulocytes with a single-laser flow cytometer. *Mutation Research.*, *371*, 283–292.

Dobrovolsky, V. N., Boctor, S. Y., Twaddle, N. C., et al. (2009). Flow cytometric detection of Pig-A mutant red blood cells using an erythroid-specific antibody: application of the method for evaluating the *in vivo* genotoxicity of methylphenidate in adolescent rats. *Environmental and Molecular Mutagenesis.*, *51*, 138-145.

Dobrovolsky, V. N., Pacheco-Martinez, M. M., McDaniel, L. P., et al. (2018). *In vivo* genotoxicity assessment of acrylamide and glycidyl methacrylate. *Food and Chemical Toxicology.*, *87*, 120-127.

Durling, L. J., Svensson, K. & Abramsson-Zetterberg, L. (2011). Furan is not genotoxic in the micronucleus assay *in vivo* or *in vitro*. *Toxicology Letters.*, *169*, 43-50.

Elhajouji, A., Vaskova, D., Downing, R., et al. (2018). Induction of *in vivo* Pig-a gene mutation but not micronuclei by 5-(2-chloroethyl)-2'-deoxyuridine, an antiviral pyrimidine nucleoside analogue. *Mutagenesis.*, *33*, 343-350.

Fenech, M. (2007). Cytokinesis-block micronucleus cytome assay. *Nature Protocols.*, *2*, 1084-1104.

Fenech, M. & Morley, A. A. (1985). Measurement of micronuclei in lymphocytes. *Mutation Research.*, *147*, 29-36.

Fernández-Bertólez, N., Costa, C., Brandão, F., et al. (2018a). Neurotoxicity assessment of oleic acid-coated iron oxide nanoparticles in SH-SY5Y cells. *Toxicology.*, 406-407, 81-91.

Fernández-Bertólez, N., Costa, C., Brandão, F., et al. (2018b). Toxicological assessment of silica-coated iron oxide nanoparticles in human Astrocytes. *Food and Chemical Toxicology.*, *118*, 13-23.

Flanagan, J. M., Howard, T. A., Mortier, N., et al. (2010). Assessment of genotoxicity associated with hydroxyurea therapy in children with sickle cell anemia. *Mutatation Research.*, *698*, 38-42.

François, M., Hochstenbach, K., Leifert, W. & Fenech, M. F. (2014). Automation of the cytokinesis-block micronucleus cytome assay by laser scanning cytometry and its potential application in radiation biodosimetry. *Biotechniques.*, *57*, 309-312.

García-Lestón, J., Roma-Torres, J., Vilares, M., et al. (2012). Genotoxic effects of occupational exposure to lead and influence of polymorphisms in genes involved in lead toxicokinetics and in DNA repair. *Environmental International.*, *43*, 29-36.

Ghanayem, B. I., Witt, K. L., Kissling, G. E., et al. (2005). Absence of acrylamide-induced genotoxicity in CYP2E1-null mice: evidence consistent with a glycidamide-mediated effect. *Mutation Research.*, *578*, 284-297.

Gradecka-Meesters, D., Palus, J., Prochazka, G., et al. (2011). Assessment of the protective effects of selected dietary anticarcinogens against DNA damage and cytogenetic effects induced by benzo[a]pyrene in C57BL/6J mice. *Food and Chemical Toxicology.*, *49*, 1674-1683.

Graupner, A., Eide, D. M., Brede, D. A., et al. (2017). Genotoxic effects of high dose rate X-ray and low dose rate gamma radiation in Apc$^{Min/+}$ mice. *Environmental and Molecular Mutagenesis.*, *58*, 560-569.

Grawé, J., Biko, J., Lorenz, R., et al. (2005). Evaluation of the reticulocyte micronucleus assay in patients treated with radioiodine for thyroid cancer. *Mutation Research.*, *583*, 12-25.

Grawé, J., Zetterberg, G. & Amnéus, H. (1992). Flow-cytometric enumeration of micronucleated polychromatic erythrocytes in mouse peripheral blood. *Cytometry*, *13*, 750–758.

Guérard, M., Koenig, J., Festag, M., et al. (2013). Assessment of the genotoxic potential of azidothymidine in the comet, micronucleus, and Pig-a assay. *Toxicology Science.*, *135*, 309-316.

Hamasaki, K., Imai, K., Hayashi, T., et al. (2007). Radiation sensitivity and genomic instability in the hematopoietic system: Frequencies of micronucleated reticulocytes in whole-body X-irradiated BALB/c and C57BL/6 mice. *Cancer Science.*, *98*, 1840-1844.

Harper, S. B., Dertinger, S. D., Bishop, M. E., et al. (2007). Flow cytometric analysis of micronuclei in peripheral blood reticulocytes III. An efficient method of monitoring chromosomal damage in the beagle dog. *Toxicology Science.*, *100*, 406–414.

Harrod, V. L., Howard, T. A., Zimmerman, S. A., et al. (2007). Quantitative analysis of Howell-Jolly bodies in children with sickle cell disease. *Experimental Hematology.*, *35*, 179-183.

Hayashi, M., Norppa, H., Sofuni, T. & Ishidate, M. Jr. (1992). Flow cytometric micronucleus test with mouse peripheral erythrocytes. *Mutagenesis.*, *7*, 257–264.

Hintzsche, H., Hemmann, U., Poth, A., et al. (2017). Fate of micronuclei and micronucleated cells. *Mutation Research.*, *771*, 85-98.

Hobbs, C. A., Chhabra, R. S., Recio, L., et al. (2012). Genotoxicity of styrene-acrylonitrile trimer in brain, liver, and blood cells of weanling F344 rats. *Environmental and Molecular Mutagenesis.*, *53*, 227-238.

Hotchkiss, C. E., Bishop, M. E., Dertinger, S. D., et al. (2008). Flow cytometric analysis of micronuclei in peripheral blood reticulocytes IV: An index of chromosomal damage in the rhesus monkey (Macaca mulatta). *Toxicology Science.*, *102*, 352–358.

Hutter, K. J. & Stöhr, M. (1982). Rapid detection of mutagen induced micronucleated erythrocytes by flow cytometry. *Histochemistry.*, *75*, 353–362.

Jain, A. K., Singh, D., Dubey, K., et al. (2019). Zinc oxide nanoparticles induced gene mutation at the HGPRT locus and cell cycle arrest associated with apoptosis in V-79 cells. *Journal of Applied Toxicology.*, *39*, 735-750.

Jiang, X., Foldbjerg, R., Miclaus, T., et al. (2013). Multi-platform genotoxicity analysis of silver nanoparticles in the model cell line CHO-K1. *Toxicology Letters.*, *222*, 55-63.

Jugan, M. L., Barillet, S., Simon-Deckers, A., et al. (2012). Titanium dioxide nanoparticles exhibit genotoxicity and impair DNA repair activity in A549 cells. *Nanotoxicology.*, *6*, 501-513.

Kiliç, G., Costa, C., Fernández-Bertólez, N., et al. (2015). *In vitro* toxicity evaluation of silica-coated iron oxide nanoparticles in human SHSY5Y neuronal cells. *Toxicology Research.*, *5*, 235-247.

Kligerman, A. D., Young, R. R., Stankowski, L. F., Jr. et al. (2015). An evaluation of 25 selected ToxCast chemicals in medium-throughput assays to detect genotoxicity. *Environmental and Molecular Mutagenesis.*, *56*, 468-476.

Kohlpoth, M., Rusche, B. & Nüsse, M. (1999). Flow cytometric measurement of micronuclei induced in a permanent fish cell line as a possible screening test for the genotoxicity of industrial waste waters. *Mutagenesis.*, *14*, 397-402.

Kotova, N., Frostne, C., Abramsson-Zetterberg, L., et al. (2015). Differences in micronucleus frequency and acrylamide adduct levels with hemoglobin between vegetarians and non-vegetarians. *European Journal of Nutrition.*, *54*, 1181-1190.

Kumar, S., Behari, J. & Sisodia, R. (2012). Influence of electromagnetic fields on reproductive system of male rats. *International Journal of Radiation Biology.*, *89*, 147-154.

Labash, C., Avlasevich, S. L., Carlson, K., et al. (2016). Mouse pig-a and micronucleus assays respond to n-ethyl-n-nitrosourea, benzo[a]pyrene, and ethyl carbamate, but not pyrene ormethyl carbamate. *Environmental and Molecular Mutagenesis.*, *57*, 28-40.

Laffon, B., Aguilera, F., Ríos-Vázquez, J., et al. (2014). Follow-up study of genotoxic effects in individuals exposed to oil from the tanker Prestige, seven years after the accident. *Mutation Research. Genetic Toxicology and Environmental Mutagenesis.*, *760*, 10-16.

Lähdetie, J. & Grawé, J. (1997). Flow cytometric analysis of micronucleus induction in rat bone marrow polychromatic erythrocytes by 1,2;3,4-

diepoxybutane, 3,4-epoxy-1-butene, and 1,2-epoxybutane-3,4-diol. *Cytometry.*, *28*, 228-235.

Le Bihanic, F., Di Bucchianico, S., Karlsson, H. L. & Dreij, K. (2016). *In vivo* micronucleus screening in zebrafish by flow cytometry. *Mutagenesis.*, *31*, 643-653.

Le Hégarat, L., Puech, L., Fessard, V., et al. (2003). Aneugenic potential of okadaic acid revealed by the micronucleus assay combined with the FISH technique in CHO-K1 cells. *Mutagenesis.*, *18*, 293–298.

Liu, L., Liu, Y., Ni, G. & Liu, S. (2009). Flow cytometric scoring of micronucleated reticulocytes as a possible high-throughput radiation biodosimeter. *Environmental and Molecular Mutagenesis.*, *51*, 215-221.

Llorente, M. T., Parra, J. M., Sánchez-Fortún, S. & Castaño, A. (2012). Cytotoxicity and genotoxicity of sewage treatment plant effluents in rainbow trout cells (RTG-2). *Water Research.*, *46*, 6351-6358.

Lukamowicz, M., Kirsch-Volders, M., Suter, W. & Elhajouji, A. (2011). *In vitro* primary human lymphocyte flow cytometry based micronucleus assay: simultaneous assessment of cell proliferation, apoptosis and MN frequency. *Mutagenesis.*, *26*, 763-770.

Lukamowicz, M., Woodward, K., Kirsch-Volders, M., et al. (2011). A flow cytometry based *in vitro* micronucleus assay in TK6 cells--validation using early stage pharmaceutical development compounds. *Environmental and Molecular Mutagenesis.*, *52*, 363-372.

MacGregor, J. T., Bishop, M. E., McNamee, J. P., et al. (2006). Flow cytometric analysis of micronuclei in peripheral blood reticulocytes: II. An efficient method of monitoring chromosomal damage in the rat. *Toxicology Science.*, *94*, 92–107.

Manjanatha, M. G., Bishop, M. E., Pearce, M. G., et al. (2014). Genotoxicity of doxorubicin in f344 rats by combining the comet assay, flow-cytometric peripheral bloodmicronucleus test, and pathway-focused gene expression profiling. *Environmental and Molecular Mutagenesis.*, *55*, 24-34.

McKeon, M., Xu, Y., Kirkland, D., et al. (2012). Cyclophosphamide and etoposide canine studies demonstrate the cross-species potential of the flow cytometric peripheral blood micronucleated reticulocyte endpoint. *Mutation Research.*, *742*, 79-83.

Morris, S. M., Dobrovolsky, V. N., Shaddock, J. G., et al. (2009). The genetic toxicology of methylphenidate hydrochloride in non-human primates. *Mutation Research.*, *673*, 59-66.

Nicolette, J., Diehl, M., Sonders, P., et al. (2010). In vitro micronucleus screening of pharmaceutical candidates by flow cytometry in Chinese hamster V79 cells. *Environmental and Molecular Mutagenesis.*, *52*, 355-362.

Nüsse, M., Beisker, W., Kramer, J., et al. (1994). Chapter 9 Measurement of micronuclei by flow cytometry. *Methods in Cell Biology.*, *42*, 149–158.

Nüsse, M. & Kramer, J. (1984). Flow cytometric analysis of micronuclei found in cells after irradiation. *Cytometry.*, *5*, 20-25.

Nüsse, M. & Marx, K. (1997). Flow cytometric analysis of micronuclei in cell cultures and human lymphocytes: advantages and disadvantages. *Mutation Research.*, *392*, 109-115.

OECD. (2016). Test No. 474: Mammalian Erythrocyte Micronucleus Test, OECD Guidelines for the Testing of Chemicals, Section 4. *OECD Publishing*. Paris.

OECD. (2016). Test No. 487: *In Vitro* Mammalian Cell Micronucleus Test, OECD Guidelines for the Testing of Chemicals, Section 4. *OECD Publishing*. Paris.

Ortiz, R., Medina, H., Cortés, E., et al. (2011). Trimethoprim-sulfamethoxazole increase micronuclei formation in peripheral blood from weanling well-nourished and malnourished rats. *Environmental and Molecular Mutagenesis.*, *52*, 673-680.

Pardini, B., Viberti, C., Naccarati, A., et al. (2017). Increased micronucleus frequency in peripheral blood lymphocytes predicts the risk of bladder cancer. *British Journal of Cancer.*, *116*, 202-210.

Paulsson, B., Grawé, J. & Törnqvist, M. (2002). Hemoglobin adducts and micronucleus frequencies in mouse and rat after acrylamide or N-methylolacrylamide treatment. *Mutation Research.*, *516*, 101-111.

Phonethepswath, S., Avlasevich, S. L., Torous, D. K., et al. (2013). Flow cytometric analysis of Pig-a gene mutation and chromosomal damage induced by procarbazine hydrochloride in CD-1 mice. *Environmental and Molecular Mutagenesis.*, *54*, 294-298.

Podrimaj-Bytyqi, A., Borovečki, A., Selimi, Q., et al. (2018). The frequencies of micronuclei, nucleoplasmic bridges and nuclear buds as biomarkers of genomic instability in patients with urothelial cell carcinoma. *Scientific Reports.*, *8*, 17873.

Rodrigues, A. (2018). Automation of the *in vitro* micronucleus assay using the ImageStream® imaging flow cytometer. *Cytometry A.*, *93*, 706-726.

Rodrigues, M. A., Beaton-Green, L. A., Kutzner, B. C. & Wilkins, R. C. (2014). Automated analysis of the cytokinesis-block micronucleus assay for radiation biodosimetry using imaging flow cytometry. *Radiation and Environmental Biophysics.*, *53*, 273-282.

Rodrigues, M. A., Beaton-Green, L. A., Wilkins, R. C. & Fenech, M. F. (2018). The potential for complete automated scoring of the cytokinesis block micronucleus cytome assay using imaging flow cytometry. *Mutation Research Genetic Toxicology and Environmental Mutagenesis.*, *836*, 53-64.

Rossnerova, A., Spatova, M., Schunck, C. & Sram, R. J. (2011). Automated scoring of lymphocyte micronuclei by the MetaSystems Metafer image cytometry system and its application in studies of human mutagen sensitivity and biodosimetry of genotoxin exposure. *Mutagenesis.*, *261*, 169-175.

Sahu, S. C., Njoroge, J., Bryce, S. M., et al. (2014). Comparative genotoxicity of nanosilver in human liver HepG2 and colon Caco2 cells evaluated by a flow cytometric *in vitro* micronucleus assay. *Journal of Applied Toxicology.*, *34*, 1226-1234.

Sahu, S. C., Njoroge, J., Bryce, S. M., et al. (2016). Flow cytometric evaluation of the contribution of ionic silver to genotoxic potential of nanosilver in human liver HepG2 and colon Caco2 cells. *Journal of Applied Toxicology.*, *36*, 521-531.

Shi, J., Krsmanovic, L., Bruce, S., et al. (2011). Assessment of genotoxicity induced by 7,12-dimethylbenz(a)anthracene or diethylnitrosamine in the Pig-a, micronucleus and Comet assays integrated into 28-day repeat dose studies. *Environmental and Molecular Mutagenesis.*, *52*, 711-720.

Smolewski, P., Ruan, Q., Vellon, L. & Darzynkiewicz Z. (2001). Micronuclei assay by laser scanning cytometry. *Cytometry.*, *45*, 19-26.

Speit, G., Zeller, J., Schmid, O., et al. (2009). Inhalation of formaldehyde does not induce systemic genotoxic effects in rats. *Mutation Research.*, *677*, 76-85.

Stankowski, L. F., Jr. Roberts, D. J., Chen, H., et al. (2011). Integration of Pig-a, micronucleus, chromosome aberration, and Comet assay endpoints in a 28-day rodent toxicity study with 4-nitroquinoline-1-oxide. *Environmental and Molecular Mutagenesis.*, *52*, 738-747.

Tometsko, A., Dertinger, S. & Torous, D. (1993b) Analysis of micronucleated cells by flow cytometry. 2. Evaluating the accuracy of high-speed scoring. *Mutation Research.*, *292*, 137–143.

Tometsko, A., Torous, D. & Dertinger, S. (1993a) Analysis of micronucleated cells by flow cytometry. 1. Achieving high resolution with a malaria model. *Mutation Research.*, *292*, 129–135.

Tometsko, A., Torous, D. & Dertinger, S. (1993c) Analysis of micronucleated cells by flow cytometry. 3. Advanced technology for detecting clastogenic activity. *Mutation Research.*, *292*, 145–153.

Tu, H., Zhang, M., Zhou, C., et al. (2015). Genotoxicity assessment of melamine in the *in vivo* Pig-a mutation assay and in a standard battery of assays. *Mutation Research: Genetic Toxicology and Environmental Mutagenesis.*, *777*, 62-77.

Valdiglesias, V., Costa, C., Kiliç, G., et al. (2013b). Neuronal cytotoxicity and genotoxicity induced by zinc oxide nanoparticles. *Environmental International.*, *55*, 92-100.

Valdiglesias, V., Costa, C., Sharma, V., et al. (2013a). Comparative study on effects of two different types of titanium dioxide nanoparticles on human neuronal cells. *Food and Chemical Toxicology.*, *57*, 352-561.

Valdiglesias, V., Laffon, B., Pásaro, E. & Méndez, J. (2011). Evaluation of okadaic acid-induced genotoxicity in human cells using the micronucleus test and γH2AX analysis. *Journal of Toxicology and Environmental Health A.*, *74*, 980-992.

Vallabani, N., Shukla, R. K., Konka, D., et al. (2014). TiO2 nanoparticles induced micronucleus formation in human liver (HepG2) cells:

comparison of conventional and flow cytometry based methods. *Molecular Cytogenetics*, 7, P79.

Witt, K. L., Livanos, E., Kissling, G. E., et al. (2008). Comparison of flow cytometry and microscopy-based methods for measuring micronucleated reticulocyte frequencies in rodents treated with nongenotoxic and genotoxic chemicals. *Mutation Research.*, *649*, 101-113.

Wang, Q., Rodrigues, M. A., Repin, M., et al. (2019). Automated triage radiation biodosimetry: integrating imaging flow cytometry with high-throughput robotics to perform the cytokinesis-block micronucleus assay. *Radiation Research.*, *191*, 342-351.

Zeiger, E., Recio, L., Fennell, T. R., et al. (2009). Investigation of the low-dose response in the *in vivo* induction of micronuclei and adducts by acrylamide. *Toxicology Science.*, *107*, 247-257.

Zhou, C., Zhang, M., Huang, P., et al. (2014). Assessment of 5-fluorouracil and 4-nitroquinoline-1-oxide *in vivo* genotoxicity with Pig-a mutation and micronucleus endpoints. *Environmental and Molecular Mutagenesis.*, *55*, 735-740.

In: Micronucleus Assay: An Overview
Editor: Robert C. Cole

ISBN: 978-1-53616-678-1
© 2020 Nova Science Publishers, Inc.

Chapter 6

IN VITRO MICRONUCLEUS ASSAY: SCOPE FOR GENOTOXICITY ASSESSMENT AND BEYOND

Abhipsa VF Debnath[1], Priti Mehta[2], PhD and Sonal Bakshi[1,*], PhD

[1]Institute of Science, [2]Institute of Pharmacy, Nirma University, Ahmedabad, Gujarat, India

ABSTRACT

Humans can become exposed to a variety of chemical substances that can have adverse biological effects. Among various types of toxicities the sub-lethal genotoxicity can have the most far reaching and severe consequences like cancer or abnormal progeny as per the cell type involved. Hence it is of great significance to identify and predict potential genotoxic agents by using laboratory markers and thus regulate and prevent exposure to cancer causing agents. The genotoxicity is majorly exerted through clastogenicity where the broken chromosomes with or without a centromere segregate separately from the main nucleus following cell

---

[*] Corresponding Author Email: sonal.bakshi@nirmauni.ac.in.

division, called micronucleus. Thus it is considered as a surrogate marker for the chromosomal breakage, spontaneous or induced due to the exposure to an agent. The search for an ideal biomarker for genotoxicity that is robust, sensitive, and objective has resulted in development and optimization of lab assays like chromosome aberration assay, micronucleus assay, comet assay, gamma H2AX assay etc. The micronucleus assay being one of the promising bioassays, we aim to discuss in detail the strengths and shortcomings along with the applications and scope. Following acute or chronic cellular exposure to various agents the sub lethal genetic damage manifest as various structural and numerical chromosomal aberrations which can be best detected and quantified at cytogenetic levels. The fate of certain structural aberrations observed at metaphase is such that these can be detected at interphase stage in the form of micronucleus. The scoring of frequency of such cells *in vitro* and in vivo allows for larger sample size and automation as compared to the conventional chromosomal aberration assay which requires skilled manpower and allows for smaller sample size. The assay can be used for bio monitoring of in vivo genotoxic exposures as well as for *in vitro* experimental exposures to assess the genotoxic potential of a candidate compound and similarly protective effect of a compound that may ameliorate the genotoxicity of a known clastogen. There are a number of confounding factors for the presence and persistence of chromosomal aberrations in the form of micronuclei that should be taken into account for better correlation of extent of actual genetic damage. This is necessary for forming guidelines regarding regulatory requirements as per the WHO and EPA when addressing the question of safety of any chemical agent. Recent modifications in the micronucleus assay techniques have expanded the scope of detecting not only the genetic damage, but also cell proliferation kinetics and differentiation. The practical feasibility of this assay makes it a significant cytogenetic tool.

Keywords: genotoxicity, biomarker, micronucleus, CBMN assay

DNA is the blue print that directs the fate of the cell. The cell is required to divide, secrete cytochemicals, undergo apoptosis or migrate from one place to another in response to an external stimulus. It can perform these classical functions only when the correct set of genes are encoded in order to respond to the environment. Hence, it is of utmost importance that the chromosomes, the genetic material carrying the 'code of life', are not damaged. However human beings are constantly exposed to various environmental agents like ionizing radiation, heavy metals, and pesticides which cause detrimental effects on the genetic material along with other cellular toxicities. The structural damage inflicted on chromosomes leading to genetic aberrations is known as genotoxicity and the causative agents are called genotoxic agents. The most hazardous outcome of genotoxic effects is cancer, since any structural damage to the DNA can potentially be the first step towards cancer initiation. It is important to understand the mechanism underlying the genetic damage as a result of exposure, the dose and the role of cellular metabolism, in order to control and prevent the consequences of genotoxicity.

Structural genetic aberrations following exposure to genotoxic agents have a spectrum of effects at the DNA as well as cytogenetic level. Depending on the mechanism of action the genotoxic compounds can be described as aneugenic or clastogenic. Aneugens are compounds that cause numerical aberrations i.e., missing or extra copy of whole or partial chromosome. Aneugenic events can give rise to aneuploidy by two mechanisms: 1) Loss of a chromosome during cell division which leads to one normal daughter cell and one with monosomy 2) Non- disjunction of a chromosome at anaphase which leads to one trisomic and other monosomic daughter cell (Parry et al., 2002). Aneuploidy of chromosome, if survived by the cell, can play an important role in carcinogenesis when acquired at somatic level and in-birth defects like Down's syndrome when at constitutional genetic level. Duplication or loss of chromosome(s) containing genes for cell division or cell cycle arrest can increase or decrease the copy number of protooncogenes or tumor suppressor genes which may drive the cells to become hyperproliferative and divide uncontrollably.

Clastogens are compounds that cause structural aberrations, mainly DNA breakage that can reflect chromatid or chromosome break when observed at metaphase of cell cycle (Figures 1 and 2). The higher extent of genetic breakage can overwhelm the DNA repair systems and increase genomic imbalance in terms of loss of genetic material, fusion of broken chromosomes leading to dicentrics, translocation, ring, acentric fragment etc., due to chromosomal instability. The genetic instability due to unrepaired DNA damage and sustained proliferation of aberrant cells predispose to cancer risk. The non-homologous end joining of broken chromosomes can lead to novel fusion gene that codes for a fusion protein that can play a role in uncontrolled cell proliferation as widely reported in leukemia and other cancers (Johansson et al., 2019). The baseline or hyper mutagen sensitivity are in fact hallmarks of cancer prone genetic conditions viz., Fanconi anaemia, Bloom syndrome, Xeroderma Pigmentosam, Ataxia Telangiectasia, etc. (Taylor, 2001).

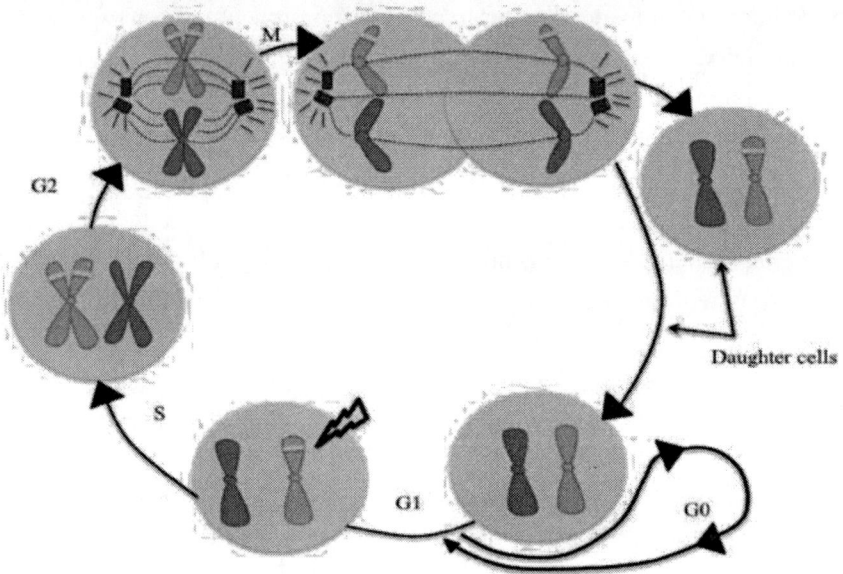

Figure 1. DNA breakage caused by a clastogen at the G1 phase of the cell cycle.

While the baseline levels of spontaneous chromosome instability due to error prone DNA replication and related factors are also responsible for cancer incidence (act of God!)[1], exposure to clastogens or aneugens increase DNA breakage and thus probability of cancer causing genetic alterations. Even in people with normal DNA repair and genetic stability, the exposure to environmental carcinogens can contribute to higher incidence of cancer. Thus a significant proportion of the cancer burden is preventable and hence should be given due attention (Doll and Peto, 2018). The measures for

---

[1] Act of God is a term used in insurance related documents to broadly indicate unpredicted reason.

preventing or controlling exposure to carcinogens and biomonitoring for probable or possible carcinogens can be taken using standard validated laboratory endpoints, such interventions can be undertaken to address the cancer incidence.

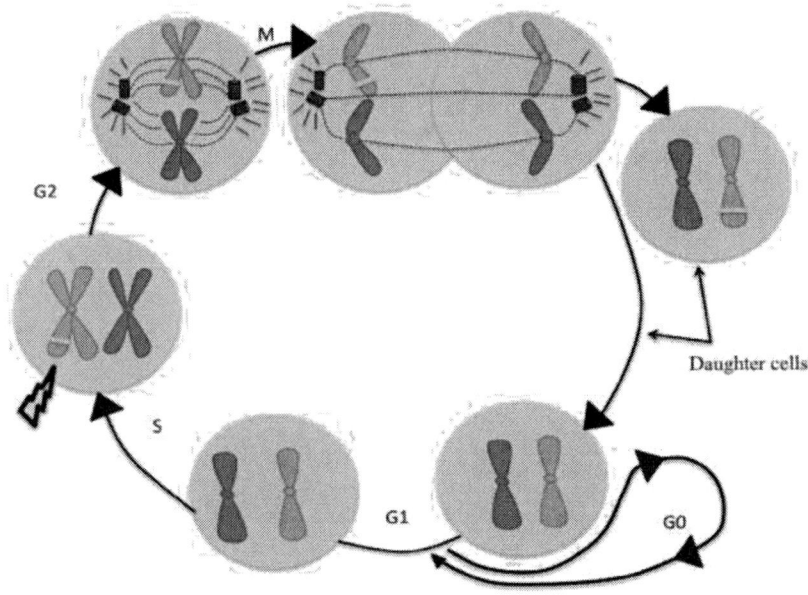

Figure 2. DNA breakage caused by a clastogen at the G2 phase of the cell cycle.

The laboratory markers for assessing DNA damage are widely reported, however, for assessing the risk of cancer *in vitro* cytogenetic approaches are most informative for predicting the outcome. The DNA damage leading to cytostasis or cell death is not relevant for cancer risk assessment, it is the cells proliferating in spite of genetic damage that raises concern for cancer initiation and promotion[2].

Acute and high levels of DNA damage can be lethal at cellular level, however, if a cell survives DNA damage and goes on to proliferate, the

---

[2] koī mere dil se pūchhe tere tīr-e-nīm-kash ko; ye k͟halish kahāñ se hotī jo jigar ke paar hotā. A famous couplet by highly acclaimed Urdu Poet Mirza Galib (1797-1869) from India is remembered here in the context of sub-lethal cellular toxicity. It can be freely translated to match the context as "the sight sharp like arrows from the bow like beautiful eyes of the loved one hurts and causes pain. If it killed, there would be no chance to feel the pain!"

resulting daughter cells are likely to carry the DNA damage in various forms. This can serve as a first hit when involving key genetic regions thus facilitating cancer onset. Micronuclei (MNi) are formed from such events either due to lagging whole or partial chromosomes also known as acentric fragments (fragments of one or both chromatids without the centromeres) both of which, upon cell division, are excluded from the main nuclei and appear as a small, separate nuclei in one or both the daughter cells (Savage, 2000). These extra nuclear bodies can be microscopically detected by DNA staining of the cells. In order to ascertain the presence of micronucleus following *in vitro* exposure, it is important to differentiate cells having undergone one cell division which is done by treating the *in vitro* cultures with a cytokinesis blocking agent such as cytochalasin-B to assess the frequency of induced chromosomal loss and chromosome breakage in binucleated cells (Fenech, 2000). The scoring of frequency of MNi also helps in distinguishing between genetic aberrations induced by aneugens and clastogens (Rosefort, Fauth and Zankl, 2004; Tinwell and Ashby, 1991). *In vitro* and *in vivo* micronucleus assay has certain advantages over metaphase chromosome aberration (CA) assay (OECD 2016) which requires skilled and experienced personnel to identify various types of chromosome aberrations and score minimum 100 to 300 metaphase cells for calculating significance as compared to controls. It is time consuming and not amenable to automation. Thus for rapid and objective scoring of genetic aberrations in larger sample size the *in vitro* micronucleus assay is recommended (Fenech, 2000). The micronucleus assay is well established for peripheral blood lymphocytes and bone marrow cells however modifications are reported or optimizing this assay using other nucleated cells as well. The detailed methodology of this assay are reported and widely followed, it is recommended to refer to the publication to avoid repetition here (Fenech, 2000).

MNi were first identified as Howell- Jolly bodies in erythrocytes by William Howell and Justin Jolly (Sears and Udden, 2012) and their pathological significance was established when Dawson *et al.* observed MNi in bone marrow cells from individuals with certain diseased conditions and with deficiency of folic acid and vitamin B12 (Dawson and Bury, 1961).

Simultaneously around the same time the role of environmental agents in causing nuclear damages was established when root tip cells exposed to ionizing radiation showed the development of MNi (Neary et al., 2016). The use of MNi as a measure of genetic damage was proposed independently by Schmid and Heddle as an alternative to chromosome aberration assay (Countryman and Heddle, 1976; Heddle et al., 1990; Schmid, 1975). Thereafter a dose response relation between ionizing radiation and the frequency of MNi in lymphocytes was shown by Fenech et al. (1986). The in vitro cytokinesis block micronucleus (CBMN) assay to measure MNi as a marker for nuclear damage in lymphocytes is a widely reported and robust assay currently (Fenech and Morley, 1985 and 1986; Bolognesi et al., 2014; Angelini et al., 2016; Cavallo et al., 2018; Dong et al., 2019).

Chromosome and/or chromatid fragments lacking a centromere (acentric fragments) and whole chromosomes with an inactive kinetochore complex which do not segregate properly at anaphase, give rise to MNi which are passed onto the daughter cells are acquired based on the type of genetic lesion (Savage, 1989; Savage, 2000). Following cytokinesis a nuclear membrane is formed around the smaller nuclei which appear as a separate entity within the cytoplasm and detected using simple nuclear stains (Fenech and Morley, 1985).

Whole lagging chromosome resulting into MNi is also reported as an age dependent phenomena (Wojda and Zie, 2007; Fenech and Morley, 1985) especially in females, loss of one X chromosome has been reported (Richard, Muleris and Dutrillaux, 1994; Hando, Nath and Tucker, 1994). Majority of the X chromosomes detected in the MNi were found to have a centromere but not a kinetochore suggesting defective kinetochore assembly in most of the X chromosomes (Hando, Nath and Tucker, 1994). The loss of Y chromosome is also reported with lack of kinetochore signals suggesting aberrant kinetochore, which showed age dependant increase (Nath, Tucker and Hando, 1995). The phenomenon of loss of whole autosomes apart from the sex chromosomes is rare (Catalán, Falck and Norppa, 2000; Falck and Catala, 2002). There are also reports of loss of chromosome 9, hypothesized to be due to the large heterochromatic block (Fauth, Scherthan and Zankl, 2000; Fauth, Scherthan and Zankl, 1998). Acrocentric chromosomes 13, 14,

15, 21, and 22 also are reported to show a higher rate of micronucleation (L Migliore, Scarpato and Falco, 1995). In a study on patients with Alzheimer's disease, centrosome positive MNi majorly involving chromosome 21 was reported to be at a higher frequency as compared to the controls (L Migliore, Testa and Scarpato, 1997; L Migliore *et al.*, 1999).

Various mechanisms are possible for the loss of whole chromosomes (Figure 3). The kinetochore complex near the centromeric region is responsible for the attachment of microtubules for the alignment of the chromosomes at the metaphase plate. Inactive or mutated kinetochore proteins can cause improper attachment of the kinetochore proteins with the spindle fibers leading to lagging behind of whole chromosomes in anaphase (Bakhoum and Genovese, 2009). The methylation status of cytosine also affects the assembly of the kinetochore proteins at the centromere and thus lead to mis-segregation of chromosomes (Gieni *et al.*, 2008). The undermethylation of cytosine leads to elongation of the pericentromeric region of the chromosomes 1, 9 and 16 hence there is improper attachment of the microtubule- kinetochore complex (Schueler and Sullivan, 2006; Gieni *et al.*, 2008; Guttenbach, 1994). Abnormal centrosome amplification, defects in mitotic spindle and mitosis checkpoint can all lead to formation of MNi containing whole chromosome (Fenech, 2000).

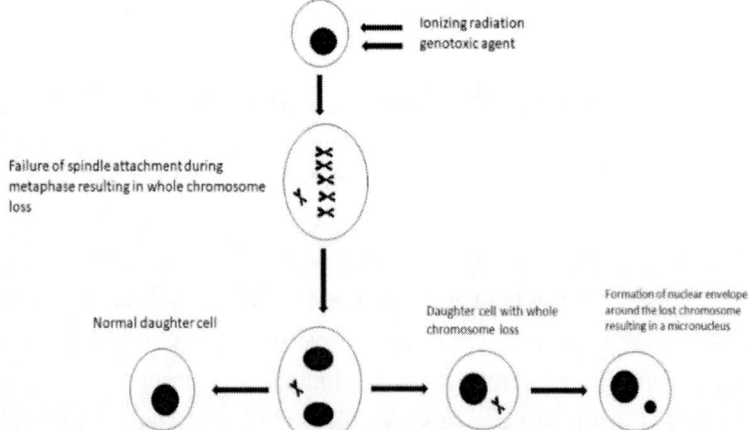

Figure 3. Mechanism of micronucleus formation due to whole chromosome loss.

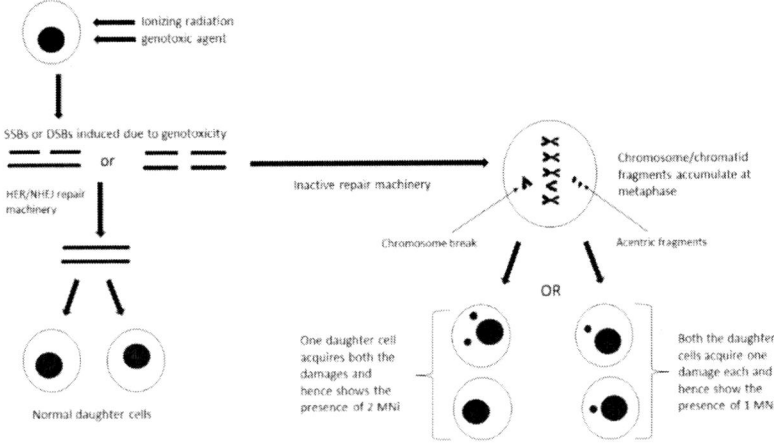

Figure 4. Mechanism of micronucleus formation due to chromosome/chromatid breaks.

Single and double strand DNA breaks remain unrepaired or get misrepaired due to defects in the repair system (Homologous Recombination Repair and Non-homologous End Joining Repair) and thus give rise to acentric chromosome or chromatid fragments forming micronuclei following mitosis (Savage, 1989; Savage, 2000) (Figure 4). The broken nucleoplasmic bridges that collapsed during the division of the daughter nuclei can also give rise to MNi (Hoffelder *et al.*, 2004). Both chromosome and chromatid type of fragments are observed in MNi with equal frequency of MNi formation. MNi also contain fragments derived from multiple segments of one or both chromatids from various chromosomes in rare instances (Savage, 2000).

In addition to these, DM (double minutes) are also reported which originate independently from two mechanisms described above. DMs are autonomously replicating, small, paired, extra nuclear bodies observed mainly in metaphase cells of human cancers. These are formed from amplified genes, showing as homogeneously staining region on chromosomes and double minutes when separated from the chromosomes stain lightly as compared to the main nucleus. These are mainly amplified oncogenes that confer proliferative advantage to the neoplastic cells

(Shimizu, 2011). The significance of these is different and has no correlation with the genotoxicity.

The genetic aberrations of certain types lead to the formation of micronucleus when the cell undergoes division. Thus detection of MNi in cultured or uncultured lymphocytes serves as a marker of genetic damage that can be quantitated. The extent of genetic damage induced by a candidate compound is compared with appropriate controls; positive, negative and vehicle control if applicable. Suboptimal DNA repair mechanisms and aberrant mitosis due to the exclusion of the damaged whole chromosome or acentric fragments during mitotic segregation at opposite poles eventually form two daughter nuclei in the subsequent step of cytokinesis. During cytokinesis the excluded DNA is entrapped inside a nuclear membrane and thus appears as a small nuclear body i.e., micronuclei. This MNi will eventually move to any one of the daughter cells. In order to detect the MNi in an *in vitro* divided cell, cytokinesis blocking agent i.e., Cytochalasin-B is added which blocks the actin monomers from polymerisation thus 'pinching off' of the cell membrane is blocked. MNi are also observed in the cells with four or more nuclei when grown in cultures with cytokinesis blocking agent. However these are disregarded for genotoxicity assays as here the cumulative effects, *in vitro* artifacts are likely to affect the persistence and number (Fenech and Morley, 1985).

This protocol for assessment of baseline and induced frequency of MNi in short term cultures of peripheral blood lymphocytes is a promising tool for genotoxicity assessment in addition to the *in vivo* rodent bone marrow micronucleated reticulocyte assay. This assay has also been adapted for use with other cell types (Odagiri, Takemoto and Fenech, 1994; Heddle *et al.*, 1990; He and Baker, 2006).

In addition to assessing the baseline and induced frequency of micronuclei *in vitro* as markers for spontaneous genetic instability and mutagen susceptibility there are other scopes for studies as listed below.

This assay can be used to distinguish between the loss of whole chromosome from the single or double chromatid fragments by including probes against centromeric DNA or antibody against kinetochore proteins [(available commercially and also from serum samples of scleroderma patients of CREST subtype (Moroi *et al.*, 1981)] (Degrassi and Tanzarella, 1988; M Fenech and Morley, 1989; Farooqi, Darroudi and Natarajan, 1993; Schuler, Rupa and Eastmond, 1997).

Non-disjunction events of a specific chromosome can be identified using complementary DNA probes in binucleated cells (Zyno, Marcon and Leopardi, 1994; Schuler, Rupa and Eastmond, 1997).

Genotoxic agents like UV radiation and certain pesticides induce DNA base lesions however the base excision repair mechanism operating in the cell cause reduced frequency when assessed. Hence inhibitors of DNA repair mechanisms like ARA-C are added which enable single strand DNA breaks also to be converted into double strand breaks, this can result in chromatid or chromosome fragments when cells replicate and form the micronuclei (Fenech and Neville, 1992; Surralles *et al.*, 1995).

The assay can also detect the presence of dicentric chromosomes which form nucleoplasmic bridges (NPB) between the two nuclei in a binucleated cell. The two centromeres of the dicentric chromosome get attached to the spindle fiber at the opposite ends and during anaphase are pulled apart forming a bridge like structure between the two nuclei (Figure 5). Usually such structures are formed following an exposure to ionising radiations (Fenech, 1997).

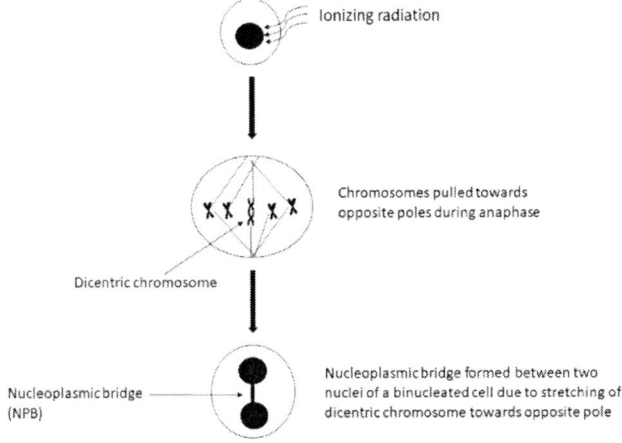

Figure 5. Mechanism of Nucleoplasmic Bridge formation due to Dicentric chromosome.

The CBMN assay is a very useful approach for measuring the nuclear division rate in dividing cell populations. In order to study the cytostatic or mitogenic effects of any drug or test compound *in vitro,* it can be assessed by comparing the number of cells having undergone first, second or third cell cycle which is reflected by the mononucleated, binucleated and multinucleated cells in cultures added with Cyt-B. The formula given by Eastmond and Tucker for calculating Nuclear Division Index (NDI) is as follows;

$$NDI = MI + 2\,MII + 3\,MIII/N$$

where MI-MIII represent the cells with 1, 2 and 4 nuclei respectively and N represents the total number of cells counted (500 viable cells are recommended to be counted) (Fenech, 2016).

Similarly, the cytotoxic effects of a test compound is assessed by adding another variant of this assay called CBMN cytome assay. Here the apoptotic as well as necrotic bodies are counted along with other viable cells and applied to the formula for Nuclear Division Cytotoxicity Index (NDCI) as follows: $Ap + Nec + MI + 2\,MII + 3\,MIII/N^*$ where Ap and Nec represent number of apoptotic and necrotic cells, MI-MIII represent the cells with 1, 2 and 4 nuclei respectively and N^* represents the total number of cells (viable and non-viable) (Fenech *et al.*, 2011; Fenech, 2016).

The impact of micronutrients in preventing DNA level damage has been recently studied. Vitamin-C, B12, folate, zinc, iron etc. play a role in the prevention of cancers (Fenech, 2018). Thus CBMN cytome assay can also unravel cytostatic effect of deficiency of any nutrients.

This assay is also employed in the screening of HPRT mutant peripheral blood lymphocytes using phosphorylated analogues of guanine or hypoxanthine which are inhibitory to normal cells and thus variants will be resistant to inhibition. The mutant cells undergo at least one cell division and hence are captured by Cyt-B treatment (Norman, Mitchell and Iwamoto, 1988).

Any candidate test compounds or known genotoxicants can induce a spectrum of genetic aberrations and not all types can be detected by any one assay. This needs to be recognized in order to develop better experimental designs and choosing the most relevant assay.

The micronucleus assay can only capture events involving loss of chromosomes and/or chromatid fragments but remain uninformative for balanced translocations, inversions, insertions etc. As discussed above, the genetic aberrations resulting in MNi *in vitro* in binucleated stage during short term cultures for 72 hrs are to be considered, which requires efficient cell division. Thus MNi detection is not the best option when studying non-dividing cells, neuronal cells, brain tissue, muscle tissue (Fenech, 1997) and similarly for cells whose cell division machinery itself has been disrupted by a genotoxic agent. If interphase death occurs extensively then the cell division process might be impeded leading to reduced detection of MNi (Savage, 2000). In such a case the measure of MNi as a marker for genotoxic insult will lead to inappropriate results. This phenomena also affects when chronic and acute exposure studies of a clastogen are required since the output of the effect in the form of a genetic aberration will be reflected in different ways for both the cases. Acute exposures when tuned at sublethal levels cause higher genetic aberrations in the form of the frequency of MNi in binucleated cells. However, chronic exposure to the same clastogen will simultaneously raise additional primary lesions and activation of the repair mechanisms leading to lower frequency of MNi in binucleated cells. Hence the comparison between chronic and acute exposure events may not be truly representative. Though not a major drawback, the assay is time consuming however since it is amenable to automation as compared to chromosome aberration assay and suitable for laboratory genetic test.

The genotoxic agents like 2,4 Diaminopurine reported to cause highly targeted breakage within the heterochromatin regions that are predominant on chromosomes 1, 9, 16 and Y are best suited for assessing the MNi using CBMN assay (Smith *et al.*, 1998).

In addition to *in vitro* MNi frequency, the target tissue can also be reflective of genotoxic damage for example studies in oral mucosa of smokers and tobacco chewers are reported with higher incidence of micronuclei linked with increased risk of oral cancer (Hans, 1992).

Occupational exposure to certain agrochemicals like organophosphates, organochemicals, pyrethroids etc. has various health risks like spontaneous abortions, congenital malformations in new-borns, sperm toxicity neurodegenerative diseases. Many are also genotoxicants and reported to linked with higher incidence of bladder cancer, pancreatic cancer, stomach cancer and liver cancer (Bosch *et al.*, 2012). Thus human biomonitoring for MNi detection can be very important for such agrochemicals.

The health concern for various novel chemical compounds are debated, however, in order to prescribe regulatory guidelines about the safety for human use the Food and Drug Administration warrants robust laboratory data for given chemical with defined dose and duration of exposure, requirement of metabolic activation if applicable including relevant positive and negative controls in the experimental design.

Table 1. Classification of candidate agents classified as per the cancer causing risk by the International Agency for Research on Cancer

Group	Category name	Definition
Group 1	The agent is carcinogenic to humans.	Sufficient evidence of carcinogenicity in humans.
Group 2	The agent is almost carcinogenic to humans.	On one hand, the degree of evidence of carcinogenicity is almost sufficient for humans and on the other hand, there is no human data but there is evidence in experimental animals.
Group 2A	The agent is probably carcinogenic to humans.	Limited evidence of carcinogenicity in humans and sufficient evidence of carcinogenicity in experimental animals. Exception: An agent may be classified only on the basis of limited evidence of carcinogenicity in humans.
Group 2B	The agent is possibly carcinogenic to humans.	Limited evidence of carcinogenicity in humans and less than sufficient evidence of carcinogenicity in experimental animals.
Group 3	The agent is not classifiable as to its carcinogenicity to humans.	Inadequate evidence of carcinogenicity in humans and inadequate or limited evidence of carcinogenicity in experimental animals.
Group 4	The agent is probably not carcinogenic to humans.	Evidence suggesting lack of carcinogenicity in humans and in experimental animals.

The micronucleus assay can be employed to generate laboratory data to weigh significance regarding carcinogenic potential as per the IARC (Table 1).

REFERENCES

Angelini Sabrina, Justo Lorenzo, Gloria Ravegnini, Giulia Sammarini and Patrizia Hrelia. 2016. "Application of the Lymphocyte Cytokinesis - Block Micronucleus Assay to Populations Exposed to Petroleum and Its Derivatives: Results from a Systematic Review and Meta-Analysis." *Mutation Research-Reviews in Mutation Research.* https://doi.org/10.1016/j.mrrev.2016.03.001.

Bakhoum Samuel F and Giulio Genovese. 2009. "Deviant Kinetochore Microtubule Dynamics Underlie Chromosomal Instability." *Current Biology.* 19 (22): 1937–42. https://doi.org/10.1016/j.cub.2009.09.055.

Bolognesi Claudia, Paolo Bruzzi, Viviana Gismondi and Samantha Volpi. 2014. "Clinical Application of Micronucleus Test: A Case-Control Study on the Prediction of Breast Cancer Risk/Susceptibility." *Plos One* 7: 1–18. https://doi.org/10.1371/journal.pone.0112354.

Bosch B, L Peralta, N Gentile, F Man, N Gorla and D Aiassa. 2012. "Micronucleus Assay as a Biomarker of Genotoxicity in the Occupational Exposure to Agrochemicals in Rural Workers." *Bulletin of Environmental Contamination and Toxicology.* 816–22. https://doi.org/10.1007/s00128-012-0589-8.

Catalán Julia, Ghita CM. Falck and Hannu Norppa. 2000. "The X Chromosome Frequently Lags behind in Female Lymphocyte Anaphase." *American Journal of Human Genetics.* 66 (2): 687–91. https://doi.org/10.1086/302769.

Cavallo Delia, Giovanna Tranfo, Cinzia Lucia, Anna Maria Fresegna, Aureliano Ciervo, Enrico Paci and Daniela Pigini. 2018. "Biomarkers of early genotoxicity and oxidative stress for occupational risk assessment of exposure to styrene in the fibreglass reinforced plastic industry." *Toxicology Letters*. https://doi.org/10.1016/j.toxlet.2018.06.006.

Countryman Paul I and John A Heddle. 1976. "The production of micronuclei from chromosome." *Mutation Research* 41: 321–31.

Dawson DW and HPR Bury. 1961. "The Significance of Howell-Jolly Bodies and Giant Metamyelocytes in Marrow Smears." *Journal of Clinical Pathology* 374–80.

Degrassi Francesca and Caterina Tanzarella. 1988. "Immunofluorescent Staining of Kinetochores in Micronuclei: A New Assay for the Detection of Aneuploidy." *Mutation Research* 203: 339–45.

Doll Richard and Richard Peto. 1981. "The Causes of Cancer: Quantitative Estimates of Avoidable Risks of Cancer in the United States Today." *Journal of National Cancer Institute* 66(6):1191-1308.

Dong Ju, Jun-qin Wang, Qin Qian, Guo-chun Li, Dong-qin Yang and Chao Jiang. 2019. "Micronucleus Assay for Monitoring the Genotoxic effects of Arsenic in Human Populations: A Systematic Review of the Literature and Meta-Analysis." *Mutation Research-Reviews in Mutation Research*. 780 (February): 1–10. https://doi.org/10.1016/j.mrrev.2019.02.002.

Falck Ghita C and Julia Catala. 2002. "Nature of Anaphase Laggards and Micronuclei in Female Cytokinesis-Blocked Lymphocytes." *Mutagenesis* 17 (2): 111–17.

Farooqi Z, F Darroudi, and AT Natarajan. 1993. "The Use of Fluorescence in Situ Hybridization for the Detection of Aneugens in Cytokinesis-Blocked Mouse Splenocytes." *Mutagenesis* 8 (4): 329–34.

Fauth E, H Scherthan and H Zankl. 1998. "Frequencies of Occurrence of All Human Chromosomes in Micronuclei from Normal and 5-Azacytidine-Treated Lymphocytes as Revealed by Chromosome Painting." *Mutagenesis* 13 (3): 235–41.

Fauth Evelyne, Harry Scherthan and Heinrich Zankl. 2000. "Chromosome Painting Reveals Specific Patterns of Chromosome Occurrence in Mitomycin C- and Diethylstilboestrol-Induced Micronuclei." *Mutagenesis* 15 (6): 459–67.

Fenech M, M Kirsch-Volders, A T Natarajan, J Surralles, J W Crott, J Parry, H Norppa, DA Eastmond, JD Tucker and P Thomas. 2011. "Molecular Mechanisms of Micronucleus, Nucleoplasmic Bridge and Nuclear Bud Formation in Mammalian and Human Cells." *Mutagenesis* 26 (1): 125–32. https://doi.org/10.1093/mutage/geq052.

Fenech M and A A Morley. 1989. "Kinetochore Detection in Micronuclei: An Alternative Method for Measuring Chromosome Loss." *Mutagenesis* 4 (2): 98–104.

Fenech M and S Neville. 1992. "Conversion of Excision-Repairable DNA Lesions to Micronuclei within one cell cycle in Human Lymphocytes." *Environmental and Molecular Mutagenesis* 36 (1992): 27–36.

Fenech Michael. 1997. "The Advantages and Disadvantages of the Cytokinesis-Block Micronucleus Method." *Mutation Research* 11–18.

Fenech Michael. 2000. "The *In Vitro* Micronucleus Technique." *Mutation Research* 455 (2000): 81–95.

Fenech Michael. 2016. "Cytokinesis-Block Micronucleus Cytome Assay." *Nature Protocols*. https://doi.org/10.1038/nprot.2007.77.

Fenech Michael. 2018. "Dietary Reference Values of Individual Micronutrients and Nutriomes for Genome Damage Prevention: current status and a road map to the future." *American Journal of Clinical Nutrition* 91 (January): 1438–54. https://doi.org/10.3945/ajcn.2010.28674D.1.

Fenech Michael and Alexander A Morley. 1985. "Measurement of Micronuclei in Lymphocytes." *Mutation Research* 147: 29–36.

Gieni Randall S, Gordon KT Chan and Michael J Hendzel. 2008. "Epigenetics Regulate Centromere Formation and Kinetochore Function." *Journal of Cellular Biochemistry* 2039: 2027–39. https://doi.org/10.1002/jcb.21767.

Guttenbach. 1994."Exclusion of Specific Human Chromosomes into Micronuclei by 5-Azacytidine Treatment of Lymphocyte Cultures." *Experimental Cell Research* 211:127-132

Hando John C, Joginder Nath and James D Tucker. 1994. "Sex Chromosomes, Micronuclei and Aging in Women." *Chromosoma* 186–92.

Hans F. 1992. "Localized formation of micronuclei in the oral mucosa and tobacco specific nitrosamines in the saliva of reverse smokers, khaini-tobacco chewers and gudakhu users." *International Journal of Cancer* 176: 172–76.

He Shuilin and Robert SU Baker. 2006. "Initiating Carcinogen, Triethylenemelamine, Induces Micronuclei in Skin Target Cells." *Environmental and Molecular Mutagenesis* 5 (1989): 1–5.

Heddle John A, A Bouch, MA Khan and JD Gingerich. 1990. "Concurrent Detection of Gene Mutations and Chromosomal Aberrations Induced in Vivo in Somatic Cells." *Mutagenesis* 5 (2): 179–84.

Hoffelder Diane R, Li Luo, Nancy A Burke, Simon C Watkins, Susanne M Gollin and William S Saunders. 2004. "Resolution of Anaphase Bridges in Cancer Cells." *Chromosoma* 389–97. https://doi.org/10.1007/s00412-004-0284-6.

Johansson Bertil, Fredrik Mertens, Tommy Schyman, Jonas Björk, Nils Mandahl and Felix Mitelman. 2019. "Most Gene Fusions in Cancer Are Stochastic Events." *Genes Chromosomes and Cancer* 58 (9): 607–11. https://doi.org/10.1002/gcc.22745.

Migliore Lucia, N Botto, R Scarpato, L Petrozzi and G Cipriani. 1999. "Preferential Occurrence of Chromosome 21 Malsegregation in Peripheral Blood Lymphocytes of Alzheimer Disease Patients." *Cytogenetics and Cell Genetics* 46: 41–46.

Migliore Lucia, R Scarpato and P Falco. 1995. "The Use of Fluorescence in Situ Hybridization with a β satellite DNA Probe for the Detection of Acrocentric Chromosomes in Vanadium-Induced Micronuclei." *Cytogenetics and Cell Genetics* 219: 215–19.

Migliore Lucia, Anna Testa and Roberto Scarpato. 1997. "Spontaneous and Induced Aneuploidy in Peripheral Blood Lymphocytes of Patients with Alzheimer's Disease." *Human Genetics* 299–305.

Moroi Yasuoki, Anthony L Hartman and Paul K Nakane 1981. "Distribution of Kinetochore (Centromere) Antigen in Mammalian Cell Nuclei." *The Journal of Cell Biology* (90): 1–6.

Nath Joginder, James D Tucker and John C Hando 1995. "Y Chromosome Aneuploidy, Micronuclei, Kinetochores and Aging in Men." *Chromosoma* 725–31.

Neary GJ, HJ Evans, SM Tonkinson and F S Williamson 1959. "The Relative Biological Efficiency of Single Doses of Fast Neutrons and Gamma-Rays on Vicia Faba Roots and the Effect of Oxygen." *International Journal of Radiation Biology* 7616 (March). https://doi.org/10.1080/09553005914550321.

Norman Amos, J Cameron Mitchell and Keisuke S Iwamoto. 1988. "A Sensitive Assay for 6-Thioguanine-Resistant Lymphocytes." *Mutation Research* 208: 17–19.

OECD 2016. "*OECD 473.*"

Odagiri Youichi, Kazuo Takemoto and Michael Fenech. 1994. "Micronucleus Induction in Cytokinesis-Blocked Mouse Bone Marrow Cells *In Vitro* following *In Vivo* Exposure to X-Irradiation and Cyclophosphamide." *Environmental and Molecular Mutagenesis* 24: 61-67.

Parry EM, JM Parry, C Corso, A Doherty, F Haddad, TF Hermine and G Johnson 2002. "Detection and Characterization of Mechanisms of Action of Aneugenic Chemicals." *Mutagenesis* 17 (6): 509–21.

Richard Florence, Martine Muleris and Bernard Dutrillaux. 1994. "The Frequency of Micronuclei with X Chromosome Increases with Age in Human Females." *Mutation Research 316*: 1–7.

Rosefort Christiane, Evelyne Fauth and Heinrich Zankl. 2004. "Micronuclei Induced by Aneugens and Clastogens in Mononucleate and Binucleate Cells Using the Cytokinesis Block Assay." *Mutagenesis* 19 (4): 277–84. https://doi.org/10.1093/mutage/geh028.

Savage John RK. 1989. "Acentric Chromosomal Fragments and Micronuclei: The Time-Displacement Factor." *Mutation Research* 225: 171–73.

Savage John RK 2000. "Deep Insight Section Micronuclei: Pitfalls and Problems." *Atlas of Genetics and Cytogenetics in Oncology and Haematology* (4): 229–33. https://doi.org/10.4267/2042/37683.

Schmid W 1975. "The Micronucleus Test." *Mutation Research* 31 (1975) 9-15.

Schueler Mary G and Beth A Sullivan 2006. "Structural and Functional Dynamics of Human Centromeric Chromatin." *The Annual Review of Genomics and Human Genetics.* https://doi.org/10.1146/annurev.genom.7.080505.115613.

Schuler M, DS Rupa, and DA Eastmond. 1997. "A Critical Evaluation of Centromeric Labelling to Distinguish Micronuclei Induced by Chromosomal Loss and Breakage *in vitro*." *Mutation Research* 392:81-95.

Sears David A and Mark M Udden. 2012. "Howell-Jolly Bodies: A Brief Historical Review." *The American Journal of the Medical Sciences* 343 (5): 407–9. https://doi.org/10.1097/MAJ.0b013e31823020d1.

Shimizu Noriaki. 2011. "Molecular Mechanisms of the Origin of Micronuclei from Extrachromosomal Elements." *Mutagenesis* 26 (1): 119–23. https://doi.org/10.1093/mutage/geq053.

Smith Leslie E, Karyn K Parks, Leslie S Hasegawa, David A Eastmond and Andrew J Grosovsky. 1998. "Targeted Breakage of Paracen-tromeric Heterochromatin Induces Chromosomal Instability." *Mutagenesis* 13 (5): 435–43.

Surralles Jordi, Noel Xamena, Amadeu Creus and Ricard Marcos. 1995. "The Suitability of the Micronucleus Assay in Human Lymphocytes as a New Biomarker of Excision Repair." *Mutation Research* 342: 43–59.

Taylor AMR. 2001. "Chromosome Instability Syndromes." *Clinical Hematology* 14 (3): 631–44. https://doi.org/10.1053/beha.2001.0158.

Tinwell H and J Ashby. 1991. "Micronucleus Morphology as a Means to Distinguish Aneugens and Clastogens in the Mouse Bone Marrow Micronucleus Assay." *Mutagenesis* 6 (3): 193–98.

WHO/IARC. 2006. *"Preamble to the IARC Monographs."* https://monographs.iarc.fr/wp-content/uploads/2018/06/CurrentPreamble.pdf %0Ahttp://monographs.iarc.fr/ENG/Preamble/CurrentPreamble.pdf.

Wojda Alina, and Ewa Zie. 2007. "Effects of Age and Gender on Micronucleus and Chromosome Nondisjunction Frequencies in

Centenarians and Younger Subjects." *Mutagenesis* 22 (3): 195–200. https://doi.org/10.1093/mutage/gem002.

Zinjo Andrea, Francesca Marcon, and Paola Leopardi. 1994. "Simultaneous Detection of X-Chromosome Loss and Non-Disjunction in Cytokinesis-Blocked Human Lymphocytes by in Situ Hybridization with a Centromeric DNA Probe ; Implications for the Human Lymphocyte *in vitro* Micronucleus Assay Using Cytochalasin B." *Mutagenesis* 9 (3): 225–32.

In: Micronucleus Assay: An Overview
Editor: Robert C. Cole

ISBN: 978-1-53616-678-1
© 2020 Nova Science Publishers, Inc.

Chapter 7

STUDY OF MICRONUCLEUS IN BONE MARROW CELLS OF MICE: A REVIEW

Mehnaz Mazumdar
Department of Zoology, M.C.D. College,
Cachar, Assam, India

ABSTRACT

Micronucleus assay (MA) is one of the most potential biomarker for genotoxicity studies. The conventional MA is a highly reliable method for detecting DNA damage in biological cells both *in vivo* and *in vitro*. The history of study of micronucleus (MN) dates back to 19th century. The advantages of MA are immense and hence make it the most followed test in pharmaceutical companies, drug discovery, and regulatory agencies. In the present review, generation of MN in the bone marrow cells of mice induced by various clastogenic chemicals, drugs, and radiation are elaborately elucidated. Bone marrow cells are easily susceptible to oxidative damage and sensitive to various clastogenic as well as aneugenic agents. Majority of the clastogenic agents produces reactive oxygen species (ROS) which attacks and damages the DNA of the bone marrow cells. In mice, the MN is best studied in the early and late maturing stages of RBCs known as the polychromatic erythrocytes (PCEs) and the normochromatic erythrocytes (NCEs). The MN appears in PCEs and NCEs as a tiny acentric fragment of a chromosome or the whole chromosome which lags behind in the cytoplasm during the anaphase of the cell division. Such bodies are easily stained and detected under microscope as small round or oval shaped structure. Sometimes they appear like single or multiple dots taking the same stain as that of the nucleus of the nucleated cells. For MN study two thousand PCEs are scored and the corresponding NCEs are screened as well.

INTRODUCTION

Cytogenetic assays are essential tools for measuring the genotoxic potential of various types of drugs, chemicals, radiation, food additives etc. Hsu (1982) advocated that occurrence of chromosomal anomalies in somatic cells induced by external chemical agents might increase the risk of various types of cancer and other diseases in the human body. Therefore, studies through cytogenetic assays become an essential part in the detection process. MA is one such important assay applied for detecting chromosomal abnormalities. It is a useful bio-marker for genotoxicity studies (Heddle et al. 1983) induced by various chemical clastogens (DNA damaging agent) and aneugens (Spindle poisons). It is routinely used in studying mammalian bone marrow cells in pharmaceutical companies and regulatory agencies for admission of new chemicals and drugs. Another huge advantage of the assay over other cytogenetic tests is its simplicity, speed, and cost-effectiveness. In such procedure animals are exposed to the test chemicals via appropriate administration route for specific time period. After the exposure period, the animals are sacrificed, bone marrow extracted, processed and the cells (PCEs) are analyzed under microscope for detection of MN. MNs can be defined as chromatin bodies seen in the cytoplasm of the bone marrow cells. They may appear big or small in size and may be one or more in numbers per cell. Innumerable research and studies were conducted on MN assay in the bone marrow cells of mice to establish toxicity of anticancer agents, general drugs, medicines, food additives etc. In addition, anti-genotoxic functions, radio-protectivity, modulatory role of various natural antioxidants and various plant extracts could be established using MN study as an end point.

DEVELOPMENT OF MN TEST

The history of identification of MN dates back to 19th century when they were first observed in the blood of cats and rats. They were named Howell-Jolly bodies in hematology studies. However, Evans et al. in 1959 were the first researchers to apply MN frequency as a quantitative measure for chromosomal abberation while studying gamma rays induced MN in kidney beans (Doherty et al. 2016). Boller and Schmid in 1970 for the first time developed a test to induce MN by alkylating agent trenimon in the RBC of Chinese hamster. The test was known as Mikrokern-test. In the early 1970s, it was clearly founded that MN production in the bone marrow cells of rodents *in vivo* is a measure of chromosomal damage exposed to mutagens. By mid 1970s Schmid and Heddle strongly laid the foundation of the basics of MN test. Schmid's MN protocol was further standardized and validated by Hayashi et al. in 1983 followed by MacGregor et al. in 1987.

In 1979, two groups of researchers observed MN appearance in foetal liver and peripheral blood cells of mice when the mothers were exposed to clastogens during later period of gestation (Hayashi 2016). Ever since MN test and studies have been evolving and developing fast. Initially, manual scoring of MN cells under microscope was widely done. The process is tiring and time consuming. Also, there is argument that the process is not sensitive enough since study of 2000 cells per animal is not good enough to screen the exact frequency of MN. Presently, the technique has been upgraded to semi-automated and automated techniques. In semi automated techniques image analysis of MN cells are done. Whereas, in automated technique flow cytometry is being applied to automatically count the MN (Cik et al. 2007). The frequency of MN is calculated from the cytometer with the help of computer program. The flow cytometric scoring of MN shows good results.

IMPORTANCE OF MA

The MA is one of the most popular and successful genotoxicity tests. In fact it has become an integral *in vivo* test in regulatory process of discovery and development of drugs. During the safety evaluation process, the regulatory agencies worldwide are in need of data determining the genotoxicity profile of the newly developed drugs. The MA assay helps in assessing the safety and efficacy of new drugs before their release into the market. United States, "FDAs centre for drug evaluation and research", Europe's, "The commission of the European Union", Japan's, "Ministry of health and welfare", and Canada's, "Health protection branch" have all incorporated he *in vivo* MN test in the battery of regulatory tests performed during screening and acceptance of any chemical and drug for human health welfare.

ADVANTAGES AND DISADVANTAGES OF MA

The MA study has its own advantages and disadvantages. The MN test is an independent test and does not rely on any other tests for authentication. It is simple, cost-effective, and can be easily conducted. Such test provides accurate data as MN can be easily detected and identified. In the MN erythrocytes, the main nucleus is extruded during erythropoiesis and only the MN is retained back as the DNA component of the cell. Such a characteristic of a MN cell can be best suited for identifying chromosomal aberration through automated techniques (image analysis or flow cytometry). By the help of automated process rapid scoring of MN becomes possible. Thousands of cells per treated animal can be scored increasing the assay's sensitivity (Doherty et al. 2016). Another important advantage of MN test is that any dividing cell stage can be considered for analysis irrespective of the karyotype. MN can be easily detected in interphase cells as well. The response of any clastogen can be detected for a longer period. The aneugens can also be detected and studied. The frequencies of MN are generally stable. No additional chemicals like colchicine or BrdU tablets are required to study MN frequency in cells.

The assay is of huge advantage since it is powerful alternative to chromosomal aberration test. Importantly, since MN arises from lagging chromosome, there is a possibility of detecting agents (genotoxins) which causes aneuploidy and which cannot be detected by chromosomal aberration assay. Some of the limitations include the possibility of pseudo-micronucleus detection in certain cases. It is not possible to identify the type of chromosomal aberration which might have caused the MN formation (Hayashi 2016).

FORMATION OF MN IN BONE MARROW CELLS OF MICE

Micronucleus is defined as an acentric fragment of a chromosome or the whole chromosome itself which lags behind in the cytoplasm during anaphase of cell division and does not get incorporated in the daughter nuclei.

Micronucleus is induced basically due to certain clastogenic or aneugenic effects of chemicals or other mutagens. The micronucleus study is conducted in the erythroblasts cells of bone marrow.

The RBC maturation takes place in the bone marrow of the animals. The first cell that can be identified as belonging to the red blood cell series is the proerythroblast. Large numbers of proerythroblasts are formed from the Colony forming unit- erythrocytes (CFU-E) stem cells. Once the proerythroblast is formed it divides multiple times, undergoes series of differentiations and maturations, eventually forming many mature red blood cells.

The first generations of cells which develop from the proerythroblast are basophilic in nature since they stain with basic dyes. The cells at that stage contain very little hemoglobin. The succeeding generation of cells becomes filled with hemoglobin, the nucleus condenses to a small size and the final remnant extruded from the cells.

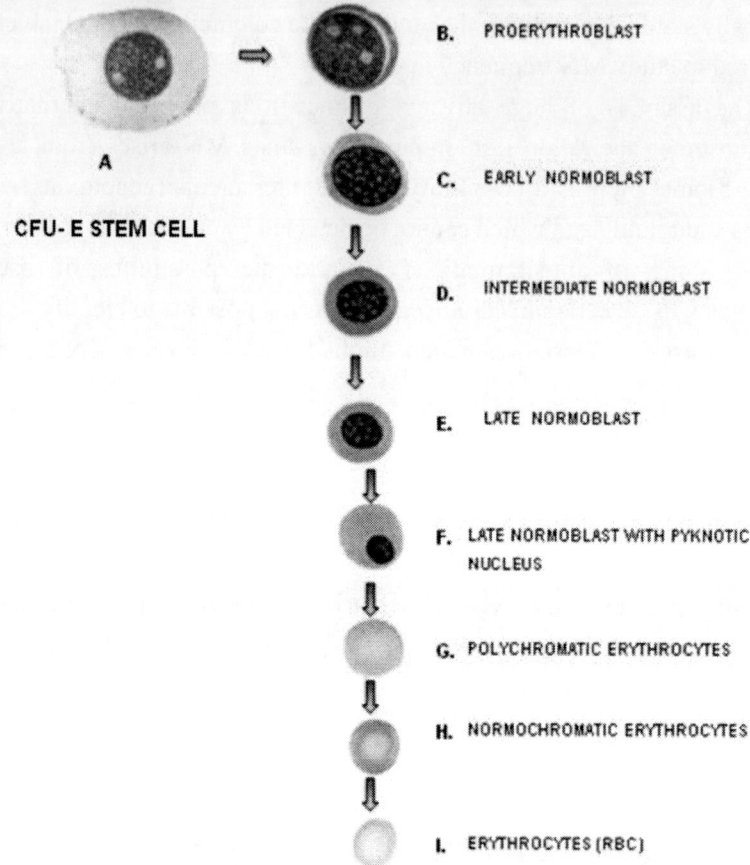

Figure 1. Pictorial depiction of different stages of a developing RBC cell (Image courtesy: Mazumdar 2010).

Hence, the changes which are evident and observed in the erythrocytes while maturation are reduction in the cell size, increase in the cytoplasmic matrix, change in the staining reaction of the cytoplasm from basophilic to acidophilic (due to decrease in the amount of RNA and DNA), reduction in the size of the nucleus and finally disappearance of the nucleus with condensed chromatin materials and the gradual acquisition of hemoglobin (Guyton and Hall 2003).

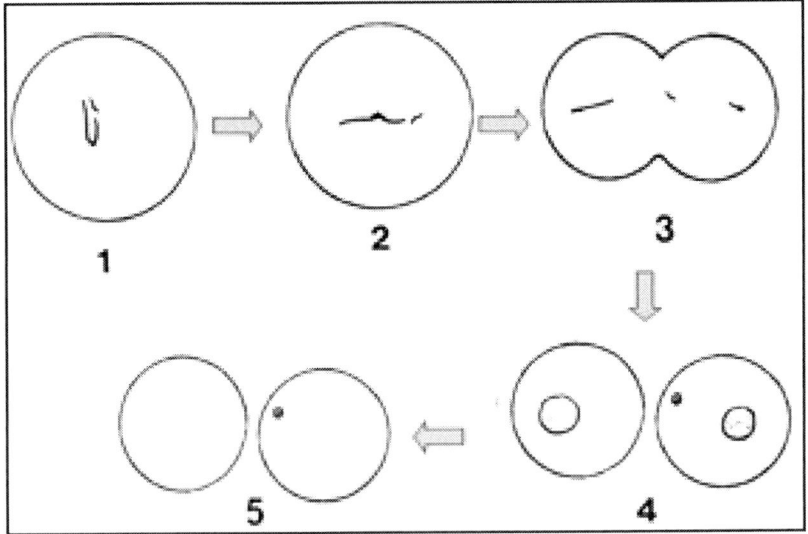

Figure 2. Schematic representation of a Micronucleus formation (1) at metaphase, (2) at anaphase, (3) at telophase, (4) the resultant micronucleus at interphase and (5) Mature cell with a micronuclei after the extrudation of the main nucleus (Image courtesy: Mazumdar 2010).

Mature RBC do not contain DNA but if chromosomal aberrations occur in the immature RBC, fragments of chromosomes may be retained in the new mature RBC. In anaphase, chromosomal fragments from the damaged chromosomes or chromosomes from abnormal segregation lag behind when the centric elements move to the spindle poles. After telophase, these lagging fragments or chromosomes may be included in the daughter nuclei. During the development, the main nuclei are expelled from the mature erythrocytes while the secondary nuclei from the fragments or lagging chromosomes are retained as "micronuclei" (Figure 2).

Polychromatic erythrocytes (PCE) are the early maturing stage of RBC taking light blue stain due to the presence of residual nucleic acids whereas normochromatic erythrocytes (NCE) are the older maturing stage of RBC and stains light reddish to transparent due to absence of nucleic acids in it (Li and Heflich 1991). 2000 polychromatic erythrocytes (PCE) and corresponding normochromatic erythrocytes (NCE) were scored per animal (Mazumdar 2010). Also, the ratio of PCE/NCE is scored to evaluate the cytotoxic effect of the chemicals tested. A PCE/NCE ratio is significantly lower than that from the solvent control indicates a decrease in the number or rate of erythrocytes maturation, suggesting cytotoxicity (Li and Heflich 1991).

PRINCIPLE AND DESCRIPTION OF MA METHOD IN MICE

The principle behind the study of MA is that animals are first exposed to the test substance via various routes which could be either, intra-peritoneal or intra-muscular, oral, or intra-venous. In the bone marrow analysis the animals are sacrificed after appropriate period of exposure to the test chemicals. There are different treatment schedules. Most studies are conducted after 24, 48 and 72 hour of exposure. After the sacrifice of the animals, the bone marrow is extracted; cellular preparations are made on glass slides, stained and analyzed for presence of MN.

Mammalian species is best suited for MA. Mice are routinely used for the purpose. In fact the major proportions of toxicological experiments are conducted using mice. They are comparatively inexpensive to maintain, economical on space, and their biology and husbandry well-understood. But importantly mice are a highly uniform group of animals which represents a perfect mammalian model in their genome size and function; cell and molecular mechanisms as well as physiology. It was observed that 99 percent of mice genes have analogues in human, as well as the genes appear in synteny in both the genomes. Another major reason using mice is that it has a vast supply of proliferating cells in the bone marrow and therefore makes it ideal for genotoxicity studies .Various strains of mice can be used for experiments provided they are healthy young animals. The weight should be ideally between 22 to 30gms. They should be maintained at 22 degree Celsius. The animals are housed in solid-walled plastic cages with mesh lids of stainless steel. In small cages, two are housed together and in large cages, four can be housed together. Males and females are kept separately and maintained at 12 hour dark and 12 hour light cycles. Standard animal feed in the form of pellets and water are provided *ad libitum* to the animals

(Mazumdar 2010). At least 5 animals per group are used for experiments and study. The mice are selected from closely inbred colonies; hence there is less probability of genetical divergence. Also, a very small size of animals (5 animals per treatment experiments) suffices the purpose of genetical response of the animals towards the tested chemical and does not vary much from one another (Mazumdar 2010).

For the experiments, two groups of animals are selected; one group serves as the control group and the other group as the treatment group. For preparation of dosage, solid test chemical has to be either dissolved, suspended or diluted with suitable solvents or vehicle substance. Aqueous solvent or vehicle is most favorable. Liquid test chemical could be directly administered or diluted prior to treatment. For dosing test chemicals should always be freshly prepared.

After appropriate time of treatment with the test chemical, the animals are sacrificed by cervical dislocation and MN slides were prepared as per Schmid protocol.

The bone marrow cells are either obtained from femur or tibia bones. The marrow are flushed out and collected with 0.9% NaCl (pre-warmed at 37degree Celsius). Bone marrow smear was prepared, air-dried and fixed with methanol for 10 minutes and stained with 5% buffered Giemsa stain (PH 7.0). The slides were mounted in DPX using cover slip (Schmidt et al. 1976). Scoring MN manually, semi-automatically or automatically is practiced.

For statistical analysis, mean values of the data are calculated by summing the observations and dividing it by their numbers. The objective is to determine whether the distribution of the responses towards different chemicals in the treated groups differs from that of the control group. The data are expressed as mean ± standard deviation (S.D.). One way analysis of variance (ANOVA) is generally used to determine the significance of genetic parameters.

Presently computerized software are available for statistical analysis. GraphPad prism is the most common software employed for such studies (Mazumdar 2010).

MN STUDIES CONDUCTED IN MICE

The MN study in mice has been useful in establishing the genotoxicity of various pharmaceutical products. It has also helped in establishing radio-protective role, anti-genotoxic effects, and anti-mutagenic activity of numerous chemicals.

Innumerable studies and evidences are available which have shown genotoxic role of various therapeutic agents, medicines, drugs, and anti-cancer agents. Researches have shown that various therapeutic drugs are harmful to the normal proliferating cells of the body. They are found to cause damage to the somatic cells as well as germ cells of the body. Such damage may cause mutagenecity leading to cancer and various other physiological consequences in human body.

There are a wide range of anticancer drugs whose genotoxicity has been established through MN study in mice. The most common of these drugs are mitomycin C, cyclophosphamide, cisplatin (Shruthi and Shenoy 2018), bleomycin, vincristin, vinblastin, etc. Studies conducted on mitomycin C in dosage 2 mg per kg body weight to balb/C mice induced significant rise in MN-PCEs after 24 and 48 hour of exposure. Cisplatin in dosage 4 mg per kg body weight also induced high frequency of MN (Mazumdar et al. 2012). Taxol, an anticancer drug synthesized from stem bark of the pacific yew tree was found to be cytotoxic and induced higher frequencies of MN-PCEs in mice after 24 and 48 hour of exposure in all the tested doses of 1.7, 1.15, and 0.6mg per kg body weight (Rabah et al. 2010). Treatment with anticancer drug, epirubicin have shown secondary malignancies in cancer survivors. Therefore, study and evaluation of this drug became important. One study had shown that epirubicin in doses 2, 4, and 6 mg/kg had induced significant level of MN in PCEs after 30 hour of exposure of the drug (Pandit and Choudhury 2011). Also, general drug like metronidazole is a known anti-infective agent used against protozoa and anaerobic bacteria. The genotoxic potential of this drug has been established by Abrevaya, et al. applying MN test in the bone marrow cells of mice. Chloroquine is an anti-malarial agent whose genotoxic potential was evaluated using MN assay and also other cytogenetic endpoints was considered for the study. Chloroquine induced significant level of MN. Vitamin C pre-treatment was found to reduce the frequency of MN (Roy et al. 2008).

Presently, modulatory role of various antioxidants on drug induced genotoxicity is extensively studied. Studies have shown that anticancer drug, doxorubicin's genotoxicity in bone marrow cells of mice can be modulated easily by oral treatment of *Aegle marmelos* prior to drug administration. There was significant reduction in the frequency of doxorubicin induced MN (Venkatesh et al. 2007). Modulatory behavior of mitomycin C by vitamin C has been well established through numerous studies. In one of our studies, it was seen that prior administration of vitamin C in doses 125, 250, and 500mg per kg body weight to the test animals significantly reduced ($P < 0.001$) the MN frequency. MN study thus indicated antioxidant potential of vitamin C and protective intervention of the same against mitomycin C (Mazumdar et al. 2011). Mitomycin C induced MN can also be modulated by a plant flavonoid quercetin (Mazumdar et al. 2011).Giri and co-workers (1998) reported the protective role of vitamin C on cisplatin induced mutagenecity in albino mice applying MA and other tests (Giri et al. 1998). Vitamin E has shows modulatory function with drug cisplatin. It was observed that Vitamin E pre-treatment for 5 days in doses 125, 250, and 500 mg per kg body weight to Balb/C mice for 5 days reduced the MN frequency induced by Cisplatin (Mazumdar et al. 2012).

In order to treat AIDS, various nucleoside analogues are being used. In one study, 7 different nucleoside analogues were studied and it was found that all of these analogues were capable of inducing significant rise in the frequency of MN-PCE (Phillips et al. 1991). Titanium dioxide nanoparticles (TiO2-NPs) are widely used in toothpaste, ointment, printing inks, rubber, plastics, automotive products, floor coverings, ceramics, mortar, whitening food, opacity agent in pharmaceutical industry, and also used as environmental decontaminant. Study conducted on TiO2-NPs by Bakare et al. have shown significant rise in MN of mice. Herbicide glyphosate used in controlling unwanted weeds could be harmful to human beings. Study on it had indicated clastogenic and cytotoxic role of the herbicide through MN analysis (Prasad et al. 2009). Heavy metals like cobalt chloride and copper chlorides were found to be genotoxic to the bone marrow cells, due to induction of higher frequency of MN in PCEs (Rasgele et al. 2012).

Radio-protectivity of drugs like sulfasalazine (SAZ) used in inflammatory bowel disease was studied using MN test on mice bone marrow cells. In the study it was observed that MN frequency induced by gamma radiation was reduced when the mice were pre-treated with various doses of SAZ (Mantena et al. 2008). Another study also suggested radio-protectivity of herbal preparation (RH3) of medicinal plant *Hippopliae rhamnoides* against gamma rays. The findings clearly establish the fact that treatment with RH 3 prior to exposure of gamma-radiation reduce MN frequency significantly (Agrawala and Goel 2002). Amalakyadi churna, a herbal preparation made from fruits of phyllanthus emblica, piper longum L, Terminalia chebula retz and roots of plumbago also showed radioprotectivity against 4 Gy-gamma rays by reduction of MN in Swiss albino mice (Reddy 2009).

MN studies in mice bone marrow cells have been extremely useful in studying anti-genotoxic effects of various herbal drugs and anti-oxidants against wide range of clastogenic drugs. Septilin, a herbal drug was found to reduce the frequency of bone marrow MN to a significant level induced by cyclophosphamide (Shruthi and Vijayalaxmi 2016). The genoprotective effects of gallic acid, which is a phenolic plant compound had been well established through MA. Cisplatin, a well-known clastogen is capable of producing high level of MN-PCE in the bone marrow cells of mice. However, gallic acid in combination with cisplatin had shown significant reduction of MN in all the tested doses (Shruthi and Shenoy 2008).

The MN test has also been useful in studying the anti-mutagenic activity of *Rhaphidophora pinnata* leaves. The ethanol extract of the leaves in doses 500, 750, and 1000kg body weight were administered to mice orally followed by drug cyclophosphamide treatment. The study has shown decreased level of MN formation in the treated mice indicating anti-mutagenic potential of the leaves against the tested drug (Masfria and Dalmunthi 2017). 5-fluorouracil is a widely used uracil analogue against wide range of cancers. It has serious clastogenic potential and myleosupression. Antioxidant chlorophyllin pre-treatment could significantly decrease the number of PCEs with MN in a dose-dependent manner (Noshy & Ramadan 2013).

CONCLUSION

The MA conducted in mice has its own advantages as well as limitations just like any other genotoxicity assessment tests. In fact it is not possible for any single assay to assess and detect genotoxicity induced by vast range of available chemicals. This is due to the different modes of actions of different chemicals. In spite of all these, the MN assay has been the most favorable short-term test used widely for its reliability. Detecting chromosomal damage by MN assay not only helps in understanding mutagenecity but also helps in identification of hazards and risks associated with various chemicals. In today's highly advanced research world, animal studies are still very important for assessing the safety level of chemicals.

REFERENCES

Abrevaya, X. C., Carballo, M. A. and Mudry, M. D. 2007. "The bone marrow micronucleus test and metronidazole genotoxicity in different strains of mice (Mus musculus)". *Genet. Mol. Bio.*, 30(4):1139 - 1143.

Agrawala, P. K. and Goel, H. C. 2002. "Protective effect of RH-3 with special reference to radiation induced micronuclei in mouse bone marrow". *Indian Journal of Experimental Biology,* 40:525 - 530.

Bakare, A. A., Udoakang, A. J., Anifowosne, A. T., Fadoju, O. M., Ogunsuyi, O. I, Alabi, O. A, Alimba, C. G. and Oyeyemi, T. T. 2016. *Journal of pollution effects and control*, 4:2.

Cik, M. and Jurzak, M. R. 2007. "Drug discovery technologies". *Comprehensive Medicinal Chemistr.,* 3: 679 - 696.

Doherty, A., Bryce, S. M., Bemis, J. C. 2016. "Chapter 6- The in vitro Micronucleus assay". *Genetic toxicology testing, a laboratory manual,* 161 - 205.

Giri, A., Khynriam, D. and Prasad, S. B. 1998. "Vitamin C mediated protection on cisplatin induced mutagenecity in mice". *Mutation research,* 421:139 - 148.

Guyton, A. C. and Hall, J. E. 2003. *Textbook of Medical Physiology.* Saunders., 10th edition.

Hayashi, M. 2016. "The micronucleus test –most widely used in vivo genotoxicity test". *Genes Environ.,* 38:18.

Heddle, J. A., Hite, W., Kirkhart, B., Mavournin, K., MacGregor, J., Newell, G. W. and Salomone, M. F. 1983. "The induction of micronuclei as a measure of genotoxicity. A report of the U.S. environmental protection agency Gene-Tox Program". *Mutat. Res.,* 123:61 - 118.

Jain, A. K. and Pandey, A. K. 2018. "Models and methods for in vitro toxicity". *In vitro toxicology,* 45 - 65.

Jena, G. B., Kaul, C. L. and Ramarao, P. 2002. "Genotoxicity testing, A regulatory requirement for drug discovery and development: Impact of ICH guidelines". *Indian journal of pharmacology,* 34:86 - 99.

Li, A. P. and Heflich, R. H. 1991. *Genetical Toxicology* by CRC Press.

Mantena, S. K., Unnikrishnan, M. K. and Devi, P. U. 2008. "Radioprotective effect of Sulfasalazine on mouse bone marrow chromosomes". *Mutagenesis,* 23(4):285 - 292.

Masfria, S. and Dalimunthe, A. 2017. "Antimutagenic activity of ethanol extract of rhaphidophora pinnata (L.f) Schott leaves on mice". *Scientia Pharmaceutica,* 85:7.

Mazumdar, M., Giri, S. and Giri, A. 2011. "Role of quercetin on mitomycin C induced genotoxicity: Analysis of micronucleus and chromosome aberrations in vivo". *Mutation Research,* 721(2):147 - 152.

Mazumdar, M., Giri, S., Singh, S., Kausar, A., Giri, A. and Sharma, G. D. 2011. "Antioxidative Potential of Vitamin C against Chemotherapeutic Agent Mitomycin C induced Genotoxicity in Somatic and Germ Cells in Mouse Test Model". *Assam. University journal of science and technology,* 7 (1):10 - 17.

Mazumdar, Mehnaz. 2010. "Protective Action of Some Natural Antioxidants against selected Antitumor Chemotherapeutic Agents: A Comparative Study in Mouse Test System using cytogenetical endpoints *in vivo*". *PhD diss.*, Assam University.

Mazumdar, M., Giri, S. and Roy, S. 2012. "Role of vitamin E-acetate on cisplatin induced genotoxicity: An in vivo analysis". *Central European Journal of Biology,* 7(2):334 - 342.

Mazumdar, M., Giri, S. and Roy, S. 2012. "Study of cytogenetical effects of chemotherapeutic agents mitomycin C and cisplatin on normal somatic and germ cells: An *in vivo* study". *International journal of pharmtech research,* 4(1):61 - 67.

Noshy, M. M. and Ramadan, H. 2013. "Evaluation of the role of chlorophyllin in modulating the in vivo clastogenecity of the anticancer drug 5-fluorouracil in mice". *Egypt. J. experimental biology,* 9(1): 123 - 131.

Pandit, R. S. and Choudhury, R. C. 2011. "Clastogenic effects of the anticancer drug Epirubicin on mouse bone marrow cells". *Biology and Medicine,* 3(5):43 - 49.

Phillips, M. D., Nascimbeni, B., Tice, R. R., Shelby, M. D.,1991. "Induction of micronuclei in mouse bone marrow cells: An evaluation of nucleoside analogues used in the treatment of AIDS". *Environ. Mol. Mutagen.,* 18(3):168 - 183.

Prasad, S., Srivastava, S, Singh, Madhulika and Shukla, Y. 2009. "Clastogenic effects of Glyphosate in bone marrow cells of swiss albino mice", *Journal of toxicology,* 2009:1 - 6.

Rabah, S. O., Ali, S. S., Alsaggaf, S. M. and Ayuob, N. N. 2010. "Acute Taxol toxicity: The effects on bonr marrow mitotic index and the histology of mice hepatocytes". *J. Appl. Anim. Res.,* 38(2010):201 - 207.

Rasgele, P. G., Kekecoglu. and Muranli F. D. G. 2013. "Induction of micronuclei in bone marrow cells by cobalt and copper chlorides". *Archives of environmental protection,* 39(1):75 - 82.

Reddy, B. U. 2009. "An Ayurvedic preparation amalakyadi churna protects against radiation induced micronuclei in mouse bone marrow". *Pharmacology on line,* 2:75 - 83.

Roy, L. D., Mazumdar, M. and Giri, S. 2008. "Effects of low dose radiation and vitamin C treatment on Chloroquine induced genotoxicity in mice". *Environmental and Molecular Mutagenesis,* 49:488 - 495.

Shruthi, S. and K. Bhasker Shenoy. 2018. "Genoprotective effects of gallic acid against Cisplatin induced genotoxicity in bone marrow cells of mice". *Toxicol. Res.,* 7:951 - 958.

Shruthi, S. and Vijayalaxmi, K. K. 2016. "Antigenotoxic effects of a polyherbal drug septilin against the genotoxicity of cyclophosphamide in mice". *Toxicology Reports,* 3(2016)563 - 571.

Schmid, W. 1976. *"The micronucleus test for cytogenetic analysis. In: Hollaender A editor. Chemical mutagens: Principles and methods for their detection".* New York: Plenum Press. 4: 31 - 53.

Venkatesh, P., Shantala, B., Jagetia, G. C., Rao, Koteshwer. and Baliga, M. S. 2007. *"Modulation of Doxorubicin induced genotoxicity by Aegle marmelos in mouse bone marrow: A Micronucleus Study".* 6(1): 42 - 53.

INDEX

A

acid, 4, 8, 9, 66, 70, 71, 72, 73, 81, 100, 103, 104, 107, 110, 115, 118, 121, 128, 156, 160
aneugenic, viii, ix, xii, 16, 24, 25, 50, 56, 63, 76, 90, 98, 102, 105, 112, 118, 125, 141, 145, 149
aneuploidy, 20, 105, 113, 125, 149
anti-cancer, 66, 154
anticancer drug(s), 154, 155, 159
antioxidant, 8, 66, 69, 70, 71, 155
apoptosis, 21, 22, 30, 34, 79, 82, 92, 102, 116, 118, 124
assessment, x, 3, 41, 51, 53, 55, 56, 62, 71, 73, 78, 79, 81, 93, 97, 101, 102, 103, 105, 108, 109, 114, 115, 118, 121, 127, 132, 137, 157
automated system, 55, 108
automation, xi, xii, 23, 98, 99, 124, 128, 135

B

base, 22, 108, 109, 110, 111, 122, 133

biomarker(s), viii, x, xi, xii, 2, 6, 7, 9, 12, 13, 16, 17, 22, 30, 33, 34, 36, 43, 45, 64, 77, 78, 79, 86, 93, 97, 98, 107, 108, 112, 120, 124, 137, 142, 145
biomonitoring, xi, 2, 7, 10, 25, 28, 32, 38, 56, 59, 76, 79, 81, 87, 89, 91, 93, 95, 98, 101, 106, 109, 110, 113, 127, 136
blood, ix, x, 5, 6, 8, 10, 11, 12, 16, 22, 23, 28, 29, 36, 52, 54, 55, 61, 62, 65, 72, 75, 81, 84, 86, 88, 89, 90, 104, 105, 106, 109, 110, 111, 112, 113, 114, 115, 116, 118, 119, 128, 132, 134, 147, 149
bone, viii, ix, xii, 16, 22, 23, 25, 36, 37, 41, 50, 53, 55, 59, 60, 61, 62, 77, 79, 109, 117, 128, 132, 145, 146, 147, 149, 152, 153, 154, 155, 156, 157, 158, 159, 160
bone marrow, viii, ix, xii, 16, 22, 23, 25, 36, 37, 41, 53, 55, 59, 60, 61, 62, 77, 79, 109, 117, 128, 132, 145, 146, 147, 149, 152, 153, 154, 155, 156, 157, 158, 159, 160

bone marrow cells, viii, xii, 77, 128, 145, 146, 147, 153, 154, 155, 156, 159, 160
buccal mucosa, 11, 13, 32, 35, 84, 87, 92, 94, 95

C

cancer, viii, xi, 1, 2, 5, 6, 7, 8, 11, 12, 18, 20, 27, 28, 32, 34, 40, 41, 56, 66, 77, 79, 80, 82, 88, 89, 99, 106, 107, 111, 115, 119, 123, 125, 126, 127, 128, 135, 136, 146, 154
carcinogenesis, 11, 51, 52, 125
carcinogenicity, 25, 51, 63, 136
carcinoma, 32, 35, 103, 120
cell culture, 27, 33, 54, 69, 83, 102, 111, 119
cell cycle, 30, 116, 125, 126, 127, 134, 139
cell death, 30, 31, 33, 35, 36, 82, 90, 127
cell division, vii, viii, xi, xii, 1, 17, 19, 28, 34, 53, 80, 98, 99, 101, 124, 125, 128, 134, 135, 146, 149
cell line(s), 20, 27, 29, 52, 54, 67, 90, 101, 102, 103, 110, 112, 117
centromere, xi, 6, 30, 90, 123, 129, 130
chemical(s), viii, ix, xi, xii, 8, 16, 17, 19, 22, 23, 24, 27, 28, 31, 32, 50, 51, 53, 57, 59, 63, 79, 80, 81, 83, 84, 86, 102, 104, 105, 109, 112, 117, 122, 123, 136, 145, 146, 148, 149, 152, 153, 154, 157
chromosomal abnormalities, 146
chromosomal instability, 27, 30, 32, 78, 90, 125
chromosome, vii, viii, ix, x, xi, xii, 1, 8, 16, 17, 18, 19, 20, 21, 22, 25, 27, 28, 29, 30, 33, 34, 37, 50, 52, 53, 56, 57, 59, 62, 76, 77, 78, 79, 92, 94, 97, 98, 101, 107, 121, 123, 124, 125, 126, 128, 129, 130, 131, 132, 133, 135, 138, 139, 140, 146, 149, 151, 158
chromosome breakage, x, 62, 77, 92, 97, 98, 128
chromosome loss, x, 56, 92, 97, 98, 130
clastogenic, vii, viii, ix, xii, 6, 15, 16, 23, 24, 50, 56, 63, 76, 86, 94, 98, 102, 105, 111, 112, 121, 125, 145, 149, 155, 156, 159
compounds, ix, 2, 50, 51, 54, 71, 83, 84, 102, 105, 106, 109, 110, 113, 118, 125, 134, 136
computerized image investigating system, 55
control group, 5, 6, 31, 83, 84, 85, 86, 87, 153
correlation, xii, 5, 56, 83, 84, 85, 100, 103, 107, 124, 132
cyclophosphamide, 8, 36, 154, 156, 160
cytochalasin-B, 3, 28, 59, 77, 99, 128, 132
cytokinesis, 3, 8, 9, 10, 18, 27, 28, 30, 31, 34, 38, 56, 62, 63, 64, 76, 78, 82, 89, 99, 115, 120, 122, 128, 129, 132
cytokinesis-block micronucleus (CBMN) assay, 6, 9, 10, 27, 28, 29, 56, 62, 63, 64, 77, 85, 86, 89, 90, 99, 108, 120, 122, 124, 134, 135
cytome, ix, 3, 12, 16, 30, 31, 33, 36, 37, 38, 41, 64, 78, 81, 90, 92, 93, 94, 114, 115, 120, 134, 139
cytometry, vii, ix, xi, 23, 36, 37, 39, 50, 54, 58, 59, 61, 62, 64, 91, 98, 100, 101, 104, 108, 110, 111, 112, 114, 115, 116, 118, 119, 120, 121, 122, 147, 148
cytoplasm, xii, 2, 21, 24, 33, 80, 82, 98, 99, 100, 101, 129, 146, 149, 150
cytotoxicity, 56, 59, 66, 102, 112, 121, 152

D

detection, viii, ix, 1, 24, 26, 27, 32, 39, 40, 50, 55, 59, 63, 77, 99, 114, 116, 132, 135, 136, 146, 149, 160
diseases, vii, viii, 1, 2, 3, 7, 13, 27, 28, 34, 35, 80, 82, 83, 136, 146
DNA, vii, viii, xii, 2, 4, 5, 6, 7, 8, 9, 10, 11, 15, 17, 18, 19, 21, 22, 23, 24, 29, 30, 31, 32, 33, 34, 36, 37, 38, 42, 51, 52, 56, 58, 60, 66, 67, 68, 69, 77, 78, 79, 81, 82, 83, 87, 88, 89, 91, 93, 100, 102, 105, 106, 107, 110, 111, 115, 117, 124, 125, 126, 127, 131, 132, 133, 134, 139, 140, 143, 145, 146, 148, 150, 151
DNA breakage, 18, 20, 56, 58, 125, 126, 127
DNA damage, xii, 2, 4, 5, 7, 10, 11, 17, 19, 22, 31, 33, 34, 36, 38, 42, 51, 52, 66, 67, 68, 79, 83, 87, 88, 91, 93, 102, 105, 106, 107, 110, 111, 115, 125, 127, 145
DNA repair, 8, 32, 79, 82, 115, 117, 125, 126, 132, 133
drug discovery, vii, ix, xii, 50, 51, 52, 54, 55, 56, 57, 60, 145, 157, 158
drug safety, 50, 53, 58
drugs, viii, ix, xii, 3, 35, 50, 51, 53, 58, 61, 69, 79, 105, 145, 146, 148, 154, 156

E

epithelial cells, x, 3, 5, 8, 10, 33, 59, 66, 69, 76, 79, 80, 81, 82, 83, 89, 92, 93
epithelium, ix, 16, 32, 53, 80, 89
erythrocytes, x, xii, 22, 23, 24, 36, 37, 38, 39, 40, 53, 57, 75, 104, 110, 113, 114, 115, 116, 117, 128, 145, 148, 149, 150, 151

exposure, viii, ix, xi, 4, 5, 11, 12, 13, 16, 27, 28, 29, 32, 38, 40, 50, 72, 76, 77, 79, 81, 84, 85, 86, 87, 88, 89, 90, 92, 93, 94, 95, 98, 103, 105, 106, 107, 109, 115, 120, 123, 125, 126, 128, 133, 135, 136, 137, 146, 152, 154, 156

F

flow cytometry, vii, ix, xi, 23, 36, 37, 39, 50, 54, 58, 59, 61, 62, 64, 98, 100, 101, 104, 108, 110, 111, 112, 114, 116, 118, 119, 120, 121, 122, 147, 148
fluorescence, 23, 62, 100, 102, 108
fluorescence *in situ* hybridization (FISH) technique, 57, 58, 59, 78, 118
food, 3, 53, 61, 102, 103, 105, 109, 110, 146, 155
formation, 3, 5, 6, 18, 19, 21, 22, 25, 30, 33, 34, 38, 39, 77, 79, 80, 82, 83, 86, 89, 98, 119, 121, 130, 131, 132, 133, 140, 149, 151, 156
fragments, viii, x, 2, 16, 17, 18, 19, 22, 60, 77, 78, 97, 98, 128, 129, 131, 132, 133, 135, 151

G

genes, 32, 83, 94, 115, 124, 125, 131, 152
genome, 2, 9, 18, 19, 20, 27, 29, 33, 105, 111, 152
genomic instability, x, 28, 30, 32, 97, 98, 107, 116, 120
genotoxicity, vi, vii, viii, ix, x, xi, xii, 1, 2, 3, 4, 7, 9, 10, 11, 13, 16, 21, 23, 24, 25, 27, 34, 38, 41, 50, 51, 52, 54, 57, 59, 60, 61, 62, 63, 64, 67, 68, 72, 76, 78, 87, 90, 97, 98, 102, 103, 105, 108,

109, 110, 111, 112, 114, 115, 116, 117, 118, 120, 121, 122, 123, 124, 125, 132, 137, 145, 146, 148, 152, 154, 155, 157, 158, 159, 160
genotoxicity testing, 10, 21, 27, 50, 61, 62, 63, 68, 76, 98, 103, 109, 158
giemsa, 3, 24, 33, 52, 81, 153
guidelines, xii, 16, 24, 33, 58, 79, 98, 124, 136, 158

H

health, x, 1, 9, 11, 27, 35, 40, 43, 44, 45, 53, 72, 73, 75, 76, 77, 88, 91, 92, 95, 97, 121, 136, 148
high-throughput, 62, 64, 101, 106, 111, 118, 122
hoechst 33342 fluorescent staining, 52
human, viii, ix, x, xi, 1, 2, 3, 4, 5, 6, 7, 8, 9, 10, 11, 16, 17, 18, 26, 27, 28, 29, 30, 31, 35, 38, 39, 40, 41, 43, 53, 56, 58, 59, 63, 65, 66, 67, 68, 69, 70, 72, 76, 77, 87, 88, 89, 90, 91, 92, 93, 94, 95, 98, 101, 102, 103, 104, 105, 106, 107, 109, 110, 111, 112, 113, 114, 115, 117, 118, 119, 120, 121, 124, 131, 136, 146, 148, 152, 154, 155

I

image, 23, 36, 55, 59, 63, 64, 91, 99, 101, 108, 113, 120, 147, 148
in situ hybridization, 57, 61, 78
in vitro, viii, ix, x, xi, xii, 10, 11, 15, 16, 17, 21, 27, 30, 36, 39, 50, 51, 52, 54, 57, 59, 60, 61, 62, 63, 64, 66, 69, 71, 72, 75, 76, 78, 79, 92, 98, 101, 102, 103, 104, 108, 110, 111, 112, 114, 118, 120, 124, 127, 128, 129, 132, 134, 135, 142, 143, 145, 158
in vitro exposure, 128

in vivo, viii, ix, x, xi, xii, 15, 16, 17, 21, 22, 24, 27, 28, 33, 36, 37, 38, 39, 40, 50, 51, 52, 54, 56, 57, 60, 62, 63, 75, 79, 92, 94, 98, 101, 104, 105, 106, 109, 110, 111, 112, 114, 121, 122, 124, 128, 132, 145, 147, 148, 158, 159
in vivo rodent MN assay, 52
induction, vii, viii, 5, 12, 15, 22, 23, 27, 36, 52, 60, 63, 95, 102, 104, 108, 111, 113, 117, 122, 156, 158
interphase, xi, 19, 80, 86, 124, 135, 148, 151
ionizing radiation, 4, 13, 23, 36, 125, 129
iron, 104, 115, 117, 134
irradiation, 4, 37, 60, 61, 106, 119

K

kinetochore, 18, 57, 64, 129, 130, 132

L

lead, x, 17, 18, 75, 77, 86, 91, 95, 99, 106, 108, 115, 126, 130, 132, 135
lesions, 18, 32, 36, 133, 135
liver, 22, 53, 58, 62, 63, 71, 72, 102, 111, 116, 120, 121, 136, 147
liver cells, 53, 102, 111
lymphocytes, 5, 6, 8, 9, 10, 11, 27, 28, 29, 31, 35, 39, 41, 56, 58, 59, 63, 68, 72, 76, 79, 80, 83, 85, 86, 87, 88, 89, 90, 91, 98, 102, 103, 107, 111, 114, 119, 128, 129, 132, 134

M

mammalian cells, 8, 25, 28, 102, 113

marrow, viii, ix, xii, 16, 22, 23, 25, 36, 37, 41, 50, 53, 55, 59, 60, 61, 62, 77, 79, 109, 117, 128, 132, 145, 146, 147, 149, 152, 153, 154, 155, 156, 157, 158, 159, 160
metaphase, xi, 33, 77, 124, 125, 128, 130, 131, 151
methodology, 27, 101, 102, 103, 105, 109, 128
mice, vi, viii, xii, 23, 53, 61, 62, 77, 105, 106, 109, 110, 111, 115, 116, 120, 145, 146, 147, 149, 152, 154, 155, 156, 157, 158, 159, 160
micronucleus (MN), v, vi, vii, viii, ix, x, xi, xii, 1, 2, 3, 4, 5, 6, 7, 8, 9, 10, 11, 12, 15, 16, 17, 18, 19, 20, 21, 22, 23, 24, 25, 26, 27, 28, 29, 30, 31, 32, 33, 34, 35, 36, 37, 38, 39, 40, 41, 49, 50, 51, 52, 53, 54, 55, 56, 57, 58, 59, 60, 61, 62, 63, 64, 75, 76, 77, 78, 79, 80, 81, 82, 83, 84, 85, 86, 87, 88, 89, 90, 91, 92, 93, 94, 97, 98, 99, 100, 101, 102, 103, 104, 105, 106, 107, 108, 109, 110, 111, 112, 113, 114, 115, 116, 117, 118, 119, 120, 121, 122, 123, 124, 128, 129, 130, 131, 132, 135, 137, 138, 139, 141, 142, 143, 145, 146, 147, 148, 149, 151, 152, 153, 154, 155, 156, 157, 158, 160
micronucleus assay, vii, viii, xi, 1, 3, 7, 10, 11, 12, 22, 24, 27, 36, 38, 39, 40, 41, 58, 59, 60, 61, 62, 76, 87, 89, 91, 92, 93, 94, 110, 111, 112, 113, 114, 115, 117, 118, 120, 124, 128, 135, 137
microscopy, xi, 2, 19, 98, 100, 101, 103, 108, 109, 122
mitosis, 17, 19, 20, 27, 30, 37, 130, 131, 132
models, vii, viii, 15, 16, 17, 36, 53, 54, 57

mucosa, 3, 11, 13, 32, 35, 40, 54, 84, 87, 89, 92, 93, 94, 95, 135, 140
mutagen, 25, 39, 64, 116, 120, 126, 132
mutation(s), 20, 21, 39, 40, 51, 52, 58, 61, 105, 111, 112, 113, 114, 116, 119, 121, 122

N

nanoparticles, 4, 9, 10, 11, 59, 61, 66, 69, 71, 72, 103, 104, 112, 115, 116, 117, 121, 155
normochromatic erythrocytes (NCE), xii, 145, 151
nuclei, vii, viii, 1, 19, 20, 22, 26, 27, 30, 35, 56, 77, 82, 99, 101, 102, 128, 129, 131, 132, 133, 134, 149, 151
nucleic acid, 100, 151
nucleus, vii, viii, xi, xiii, 1, 2, 16, 18, 19, 20, 21, 33, 34, 82, 99, 100, 101, 123, 131, 146, 148, 149, 150, 151

O

occupational genotoxicity, 2
occupational toxicology, vii, x, 75, 76, 77, 87
Organization for Economic Cooperation and Development (OECD), 24, 27, 29, 41, 79, 93, 98, 119, 128, 141
oxidative damage, viii, xii, 13, 145
oxidative stress, 4, 11, 68, 71, 137
oxide nanoparticles, 104, 115, 116, 117, 121

P

peripheral blood, x, 5, 6, 8, 11, 12, 22, 23, 36, 54, 55, 61, 62, 75, 86, 88, 89, 105, 106, 109, 110, 111, 112, 113,

115, 116, 118, 119, 128, 132, 134, 147
pharmaceutical(s), xii, 21, 51, 62, 106, 109, 118, 119, 145, 146, 154, 155
polychromatic erythrocytes (PCE), xii, 24, 115, 117, 145, 151, 155, 156
polymorphisms, 39, 83, 90, 94, 106, 107, 115
population, ix, 4, 5, 6, 16, 17, 35, 64, 86, 88, 89, 106, 108
preparation, iv, 26, 31, 55, 63, 153, 156, 159
project, 9, 10, 16, 30, 31, 38, 80, 88, 89, 91

R

radiation, viii, xi, xii, 3, 4, 13, 23, 36, 37, 56, 62, 64, 79, 83, 92, 98, 101, 106, 107, 115, 116, 117, 118, 120, 122, 125, 129, 133, 141, 145, 146, 156, 157, 159, 160
repair, 8, 32, 79, 82, 84, 85, 89, 115, 117, 125, 126, 131, 132, 133, 135
response, 6, 23, 36, 37, 71, 79, 102, 105, 106, 112, 122, 124, 129, 148, 153
risk(s), vii, viii, xi, 1, 2, 5, 6, 8, 28, 29, 32, 39, 40, 41, 53, 56, 72, 73, 77, 79, 83, 85, 88, 89, 90, 98, 99, 101, 102, 103, 109, 111, 119, 126, 127, 135, 136, 137, 146, 157
rodents, 24, 53, 63, 79, 84, 105, 122, 147

S

safety, ix, xii, 50, 53, 55, 57, 103, 106, 109, 124, 136, 148, 157
safety reports, 51
sensitivity, 29, 30, 34, 40, 53, 79, 85, 116, 120, 126, 148

species, xii, 25, 55, 105, 118, 145, 152
spindle, 8, 18, 19, 59, 98, 105, 113, 130, 133, 151
structure, vii, x, xiii, 17, 19, 20, 50, 55, 57, 80, 82, 133, 146

T

techniques, ix, xii, 33, 50, 51, 52, 53, 54, 55, 78, 81, 100, 124, 147, 148
testing, 10, 21, 25, 27, 50, 60, 61, 62, 63, 68, 76, 98, 102, 103, 109, 158
the flow cytometric (FCM) analysis, 54
tissue, viii, x, 3, 15, 25, 33, 36, 41, 53, 55, 57, 67, 76, 77, 80, 81, 110, 135
toxicity, x, 23, 27, 54, 60, 63, 66, 67, 71, 72, 76, 77, 117, 121, 127, 136, 146, 158, 159
toxicology, vii, x, 8, 9, 17, 27, 43, 51, 52, 64, 75, 76, 77, 87, 119, 158, 159
treatment, vii, viii, 1, 6, 12, 26, 27, 28, 37, 68, 69, 76, 79, 95, 100, 103, 104, 105, 106, 118, 119, 134, 152, 153, 155, 156, 159, 160

V

validation, ix, 27, 31, 50, 57, 80, 91, 118
valuation, 52, 80, 100, 103, 105, 117, 148

W

workers, x, 4, 5, 7, 11, 12, 75, 76, 77, 83, 84, 85, 86, 87, 88, 90, 92, 93, 95, 106, 108, 137, 155
workplace, x, 75, 77

Related Nova Publications

POLYUNSATURATED FATTY ACIDS (PUFAs): FOOD SOURCES, HEALTH EFFECTS AND SIGNIFICANCE IN BIOCHEMISTRY

EDITOR: Angel Catalá, Ph.D.

SERIES: Biochemistry Research Trends

BOOK DESCRIPTION: This book presents an overview of polyunsaturated fatty acids (pufas): food sources, health effects and significance in biochemistry. The topics analyzed cover a broad spectrum of polyunsaturated fatty acids and present new information in this area of research.

HARDCOVER ISBN: 978-1-53613-572-5
RETAIL PRICE: $195

GLYCOSYLPHOSPHATIDYLINOSITOL-ANCHORED PROTEINS AND THEIR RELEASE FROM CELLS: FROM PHENOMENON TO MEANING

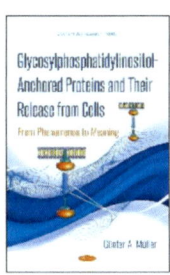

AUTHOR: Günter A. Müller

SERIES: Biochemistry Research Trends

BOOK DESCRIPTION: The book does not only represent a state-of-the-art compendium about the biology and (patho)physiology of GPI-AP, in general, and their cellular release including potential biomedical and biotechnological applications, in particular, but also introduces a novel concept for molecular life science research.

HARDCOVER ISBN: 978-1-53613-966-2
RETAIL PRICE: $230

To see a complete list of Nova publications, please visit our website at www.novapublishers.com

Related Nova Publications

LIPID RAFTS: PROPERTIES AND ROLE IN SIGNALING

EDITORS: Nils Thomas and Sten Jonathan

SERIES: Biochemistry Research Trends

BOOK DESCRIPTION: In *Lipid Rafts: Properties and Role in Signaling*, the authors summarize their observations that receptor preassembly is required for biological function; the interaction between receptor chains requires both the presence of Jak1 and their co-nanolocalization within lipid raft; a sequence-supported structural analysis of Janus kinases that suggests a significant influence of phospholipids on Janus kinase function; and critical observations made by others.

SOFTCOVER ISBN: 978-1-53613-624-1
RETAIL PRICE: $95

TRYPSIN: ANATOMY, BIOLOGICAL PROPERTIES AND APPLICATIONS

EDITORS: Diana Faas and Joshua Holder

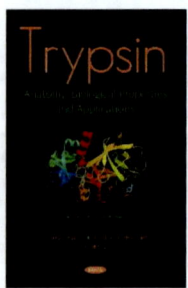

SERIES: Biochemistry Research Trends

BOOK DESCRIPTION: Trypsin, the protease with well-defined specificity, offers a great potential as a biocatalyst in numerous biomedical and industrial applications. In this collection, the authors discuss preparation and performance of trypsin immobilized on polysaccharide-based carriers.

SOFTCOVER ISBN: 978-1-53613-670-8
RETAIL PRICE: $82

To see a complete list of Nova publications, please visit our website at www.novapublishers.com